CAMBRIDGE PHYSICAL SERIES

SOUND

AN ELEMENTARY TEXT-BOOK
FOR SCHOOLS AND COLLEGES

SOUND

AN ELEMENTARY TEXT-BOOK
FOR SCHOOLS AND COLLEGES

by

J. W. CAPSTICK, O.B.E., M.A. (Camb.), D.Sc. (Vict.)

Fellow of Trinity College, Cambridge

Cambridge:
at the University Press
1922

CAMBRIDGE UNIVERSITY PRESS
Cambridge, New York, Melbourne, Madrid, Cape Town,
Singapore, São Paulo, Delhi, Mexico City

Cambridge University Press
The Edinburgh Building, Cambridge CB2 8RU, UK

Published in the United States of America by Cambridge University Press, New York

www.cambridge.org
Information on this title: www.cambridge.org/9781107674585

© Cambridge University Press 1922

This publication is in copyright. Subject to statutory exception
and to the provisions of relevant collective licensing agreements,
no reproduction of any part may take place without the written
permission of Cambridge University Press.

First Edition 1913
Second Edition 1922
First published 1922
First paperback edition 2013

A catalogue record for this publication is available from the British Library

ISBN 978-1-107-67458-5 Paperback

Cambridge University Press has no responsibility for the persistence or
accuracy of URLs for external or third-party internet websites referred to in
this publication, and does not guarantee that any content on such websites is,
or will remain, accurate or appropriate.

PREFACE

THIS book is intended primarily for students of Physics, but enough has been included of Helmholtz's Theory of Consonance to make it adequate also for students of Music. Those who wish to limit themselves to the more purely physical parts of the subject may omit everything after § 263, with the exception of §§ 279 and 280. As the later chapters will have little interest for readers with no acquaintance with Music, it has not been thought necessary to define all the musical terms that are used. Chapter XVI includes many details of the construction and use of musical instruments that cannot be regarded as necessary to be known either by students of Physics or of Music. The greater part of the chapter is intended mainly for those who play in orchestras, and who wish to know something of the principles underlying the construction of their instruments.

Certain sections of the book are marked with an asterisk to indicate that they are of a rather higher standard than the remaining sections. They can be omitted by readers who wish to acquire only an elementary knowledge of the subject, but in most cases the result reached should be known, even though the proof of that result is not read.

The writer of an elementary Text-Book on Sound suffers from the disadvantage that many of the most ordinary phenomena cannot be explained adequately without the use of mathematics of a somewhat advanced type. Consequently the argument cannot be presented in a logically continuous form, and the reader is frequently called on to accept statements the reason for which cannot be given. The best the writer can do is to endeavour to provide experimental evidence or arguments from analogy, when the theoretical explanation passes beyond the limits of the mathematical attainments that can be assumed to be possessed by his readers. I hope that in the present work I have at least made it clear where such gaps occur, and have not left the reader in doubt as to whether any statement is to be taken as a deduction from what precedes, or is to be accepted without proof. My aim has been to make the least possible use of mathematical methods, in order that the reader may not be led to evade the mental effort required for the appreciation of the physical connexion between the phenomena described. It is hoped that the book may serve as a useful introduction to the more analytical treatises, such as those of Prof. E. H. Barton and Lord Rayleigh.

I am greatly indebted to Mr T. G. Bedford, the general editor of the Cambridge Physical Series, for the valuable help he has given me throughout the preparation of the book. I have also to express my thanks to Prof. E. H. Barton for permission to make certain extracts from his Text-Book of Sound, to Mr D. J. Blaikley for much information concerning musical instruments, and to the authorities of the Universities of Cambridge, London and Dublin and the National University of Ireland for permission to make use of questions set in the examinations of those Universities.

J. W. C.

November, 1913.

PREFACE TO SECOND EDITION

NO change has been made in what was included in the First Edition, except by the correction of a few small inaccuracies and the removal of ambiguities.

An additional Chapter has been added giving an outline of some of the more important applications of Acoustics to military operations in the recent War.

J. W. C.

May 4th, 1921.

CONTENTS

CHAP.		PAGE
I	Nature of Sound	1
II	Elasticity and Vibrations	13
III	Transverse Waves	35
IV	Longitudinal Waves	57
V	Velocity of Longitudinal Waves	79
VI	Reflection and Refraction. Doppler's Principle	90
VII	Interference. Beats. Combination Tones	109
VIII	Resonance and Forced Vibrations	128
IX	Quality of Musical Notes	145
X	Organ Pipes	165
XI	Rods. Plates. Bells	181
XII	Acoustical Measurements	196
XIII	The Phonograph, Microphone and Telephone	216
XIV	Consonance	224
XV	Definition of Intervals. Scales. Temperament	239
XVI	Musical Instruments	250
XVII	Application of Acoustical Principles to Military Purposes	274
	Questions	281
	Answers to Questions	296
	Index	298

CHAPTER I

NATURE OF SOUND

1. Meanings of the word Sound. The word *Sound* has two meanings in everyday life. When we say we hear a sound, we refer to the sensation. When we say that sound travels faster with the wind than against it, we refer to some physical phenomenon external to ourselves. The two meanings of the word rarely lead to ambiguity, and no attempt will be made to distinguish the two ideas by different expressions.

2. Velocity of Sound. Many of the facts relating to sound can be readily deduced by observation of what is going on around us, without the need for special appliances.

It is a matter of common observation that sound travels much more slowly than light. The flash of a gun a mile away is seen about 5 seconds before the report is heard, and there is often a considerable interval between a flash of lightning and the resulting thunder-clap. The methods of measuring the velocity of sound will be discussed later. It will then be seen that the velocity in air at ordinary temperatures is about 1100 ft. per second, and knowing this it is easy to form an estimate of the distance of a thunderstorm, by observing the interval between the lightning and the thunder. The velocity of light is so great compared with that of sound, that for the purpose of this observation the time taken by the light in travelling from the electric discharge to the observer may be neglected.

3. Medium by which Sound is carried. If sound is carried from its source to the observer by material substances, it is clear that air must be one of these substances; for when a rocket explodes in the air, or an aeroplane passes by, the sound is heard, though there is no material substance between the source of sound and the observer, except the air.

It might be argued that we can also *see* the rocket explode, and it is obvious that the light by which we see it does not need the presence of air for its propagation, for we can also see the stars across space, which we believe to be devoid of air.

A simple experiment, however, will shew that air or some other form of matter is essential for the conduction of sound.

4. Experiment with an exhausted bell-jar. Place an electric bell or a small alarm clock on the plate of an air-pump, and cover it with a bell-jar. If the air is now pumped out, the sound heard will be somewhat weaker, shewing that the air took at least some part in the propagation of the sound. Next repeat the experiment, but place a pad of soft felt under the bell, or suspend it by india-rubber cords, and it will be found that now the sound is scarcely audible, when the air is pumped out.

From this it appears reasonable to conclude that the ether cannot conduct sound, but that the air can. The sound cannot be made to disappear entirely, for the bell must be supported by something, and all material substances carry sound, though with varying degrees of facility. When the clock rested directly on the metal plate of the pump, the sound was conveyed quite readily to the air outside, whilst the pad of felt cut it off almost entirely.

5. Propagation of Sound in Solids. Experiments to shew the propagation of sound through solids are easily devised. Lay a watch on one end of a long rod of wood or metal. The ticking of the watch will be heard when the ear is placed near the other end of the rod, but not when it is some distance away; from which we conclude that the sound heard in this case is conducted through the rod, and not directly through the air. This experiment must not be taken as proving that the material of the rod is a better conductor of sound than air. The sound communicated directly to the air by the watch spreads out freely in all directions, and rapidly grows weaker as it travels farther from the watch, whereas little of that which passes into the rod emerges until it reaches the end.

6. Velocity of Sound in different media. A similar experiment will shew that sound travels with different velocities in different media. Stand near an iron rail, whilst someone strikes the rail a few hundred yards away. Two sounds will be heard, one carried by the iron, and one by the air. It will be noticed that the sound which arrives first becomes louder if the ear is placed close to the rail, whilst the other is not affected by the change of position, so that we may conclude that sound travels more quickly through iron than through air. The velocity is in fact about 16 times as great in iron as in air. Sound is also carried readily by liquids. A person swimming under water can hear sounds made on the bank, and if two stones are struck together under water, a sound is heard in the air above the water.

7. Musical Notes and Noises. Sounds are usually divided into two classes, musical notes and noises, though there is no sharp distinction between the two. The essential difference is that musical notes have a recognizable pitch, whilst noises have not; but few noises are entirely devoid of musical pitch, and few musical notes are devoid of unmusical noise. Drop two pieces of wood of different shape on the floor. What is heard would be classed as noise, yet there would be a difference between the sounds made by the two pieces, and a person with a trained musical ear would be able to distinguish between them from the admixture of musical note with the noise. Again, we can generally by close attention detect some slight hissing of wind, or scraping, in the sound of a musical instrument.

The ear is in general able to distinguish musical notes of different pitches sounded together, but if the notes are very numerous and close together in pitch, they cease to be distinguishable, and the resultant sound is described as noise. It is probable that noise is usually such a mixture, where no one of the constituent notes is so prominent as to give a definite sensation of pitch.

8. Sound due to Vibrations. A little consideration shews that sound always takes its rise from some body which is vibrating. The ordinary method of making a tuning-fork give out its note by striking the prongs shews that what is required is to make the prongs vibrate. Look closely at them when the fork is sounding, and from the haziness of their outline it will be seen that they are vibrating. Touch them very lightly with the fingers, and the vibrations will be felt. Press the fingers more firmly on them so as to stop the vibrations, and the sound will cease. Similarly it is easily seen that a violin or harp produces sound only when a string is vibrating.

The vibrations which give rise to the sound are not so obvious in the case of a wind instrument, such as a flute or a trombone, for here it is the column of air which vibrates, and not the material of the tube. In the case of a large organ-pipe the air can be felt to be vibrating, if the hand is held near the mouth, and in other cases where the vibrations are too small to be felt, they can be detected by appropriate methods. If for instance a small paper tray with a little sand on it is lowered by a thread into an open organ-pipe with a glass panel in its side, the sand will be seen to dance about, as soon as the pipe is made to sound.

9. Sound is a Wave Motion. We have seen then that sound takes time to travel from the vibrating body to the hearer, and that it requires some material substance to carry it. It is clear that when the sound is travelling through the air it does not consist of a bodily transference of the air. If one stands in front of a trombone that is being blown loudly, no blast of air is felt, nor does a gale of wind spread outwards from a cannon when it is fired. These facts, coupled with the observation that sound always takes its rise from some body which is vibrating, suggest that what travels through the air is a series of waves of some kind.

This conjecture is consistent with all the observed properties of sound, and is in particular supported by the existence of interference, that is, by the quenching of sound at certain points by the superposition of another sound. Hold a vibrating tuning-fork to the ear and turn it round slowly with the fingers. In certain positions the sound will be almost inaudible. Whilst it is in one of these positions, get someone to slip a paper tube over one of the prongs, being careful not to touch the prong and stop its vibrations. The sound will be heard to swell out again. This experiment shews that in certain positions the sound coming from one of the prongs is able to neutralize that coming from the other, and this clearly could not happen, if sound were merely a current of air.

The phenomenon will be discussed more fully in a later chapter. It is mentioned here as being the kind of evidence on which we rely for our belief that sound consists of wave motion.

10. Characteristics of a Musical Note. Two musical sounds may differ from one another in three, and only three, ways. They may differ in loudness, in pitch, and in quality. By quality we mean the characteristic which enables the ear to distinguish, for instance, between a note played on a flute and a note of the same pitch and loudness played on a violin.

We shall discuss very briefly the way in which these characteristics of a musical note are related to the features of the vibrations that give rise to the note. What will be said must not however be regarded as a proof of the relationships, but rather as a synopsis of what will be more fully proved in the succeeding chapters.

11. Loudness. Strike a tuning-fork strongly, so as to make it give out as loud a note as possible. The extent of vibration of the prongs can be seen from the extent of the haziness at their ends. It will be noticed that this amplitude gradually diminishes, and that at the same time the sound grows weaker. Thus it appears that the greater the extent of vibration of the fork, the louder is the sound that it gives out. A similar conclusion may be drawn from the vibrations of a stretched string, such as that of a harp. The farther it is drawn aside before being let go, the greater will be the amplitude of its vibrations, and the louder the sound it will give out.

It is reasonable to suppose that the more widely the prongs of a fork vibrate, the greater will be the amplitude of the air waves sent out, and we may assume that the loudness of the sound depends in some way on the amplitude of the air waves—the greater the amplitude, the louder being the sound.

We need not discuss here the exact nature of the air waves, nor the exact meaning of the term *amplitude* as applied to them. The analogy of the familiar waves on water will serve our present purpose, though these differ in important respects from sound waves. In the case of water waves we mean by amplitude the height of a crest above the mean level of the water.

In Fig. 1 *AB* is the amplitude of the upper wave, and the two waves differ in no other respect than in amplitude. They represent, in a way that will be explained later, two sounds which have the same pitch and quality, but differ in loudness.

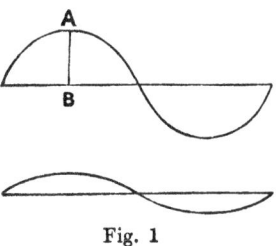

Fig. 1

12. Pitch. A simple experiment for finding what feature of the vibrations of a body characterizes the pitch of the note emitted may be made with Savart's Toothed Wheel. A metal disc has saw-teeth round its circumference, and can be rotated at various rates by a handle or a small electric motor. Turn the wheel at a gradually increasing rate, whilst a card is held against the teeth. At first separate taps are heard, as the card drops from tooth to tooth, but as the rate of rotation increases, the taps blend into a harsh note, whose pitch rises, as the wheel is made to rotate more rapidly. In this experiment the card is made to vibrate by the teeth, and the number of vibrations it makes per second depends on the rate of rotation of the wheel; whence we conclude that the pitch of a note is determined by the number of vibrations per second of the body which gives out the note.

The apparatus could be used to find how many vibrations per second correspond to any given note, such as that of a particular tuning-fork. Turn the wheel at such a rate that the card gives out the same note as the fork, and count the number of turns made by the wheel in a minute. Suppose there are 100 teeth on the wheel, and it is necessary to turn it 120 times in a minute to give a note of the same pitch as that of the fork. Then the number of taps is evidently 200 per second, and this is the number of vibrations per second required to produce the note in question. This method, however, is not capable of any great accuracy.

We shall shew in a later chapter that, when the sound travels from the vibrating body to the ear through air or other medium, the number of waves that reach the ear per second is equal to the number of vibrations per second of the vibrating body; that waves of all lengths travel with the same velocity in any one medium such as air, and that therefore the difference between a sound of high pitch and one of low pitch may be said to be that the wave-length of the former is less than the wave-length of the latter.

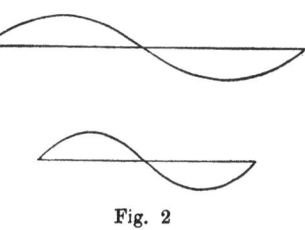

Fig. 2

Fig. 2 represents two waves of the same amplitude but of different wavelengths. The lower curve corresponds to a higher note than the upper.

13. Limits of Audibility. It should be mentioned here that a musical note is heard only when the number of vibrations per second is between certain limits, which are different for different persons.

If the vibrations are slower than about 30 per second they do not blend into a note, but are heard separately. Even when they are somewhat above 30 per second, they do not give the sensation of a note, unless they are of the kind known as Simple Harmonic Vibrations, to be described in the next chapter.

When the vibrations are very rapid, they cease to produce any impression on the ear. The sensitiveness of the ear to high notes falls off with advancing age. Children can generally hear notes with 20,000 vibrations per second, elderly people cannot generally hear anything above 15,000.

We conclude then that within these limits the pitch of a note depends on the number of vibrations executed per second by the body which gives out the sound. This number is called the *Frequency* of the vibrations.

8 NATURE OF SOUND [CH. I

14. Intervals. It is often required to express numerically the relationship between the pitches of two notes, and we must consider what is the most convenient method of measuring the interval between the notes. In deciding what method to use the chief fact of which we have to take account is the existence of certain intervals such as the octave, the fifth, etc., which give a constant and recognizable mental impression at all parts of the musical scale. A fifth, for instance, is the same interval to the ear, whether it be low in the bass or high in the treble, and whatever measure we adopt for an interval, it must be such as will give the same numerical value at any part of the scale to what the ear judges to be the same interval. We must first then find experimentally what feature of the relationship between the frequencies of two notes forming such an interval as a fifth is constant. This is readily done by the use of the Disc Siren.

15. The Disc Siren. A circular disc of cardboard or metal is mounted on an axle, so that it can be rotated rapidly. The disc has several circles of holes pierced through it, and a jet of air from the mouth or bellows can be directed on one of the circles by a narrow glass tube. Whenever a hole comes in front of the tube a puff of air passes through the disc, so that we have a rapid succession of puffs, which will be found to blend into a musical note, and the pitch of the note will depend on the rate of rotation of the disc, and on the number of holes in the circle.

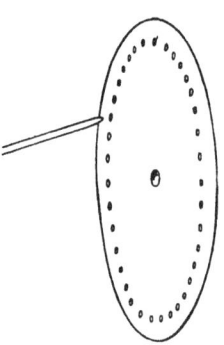

Fig. 8

Let us suppose there are four circles of holes with 40, 50, 60, and 80 holes in the respective circles. Whilst the disc is turned at a steady rate, direct the jet of air first at the 80 circle, and then at the 40 circle. The interval between the two notes heard will be recognized as being an octave. Turn the disc more quickly, and both notes will rise in pitch, but the interval between them will still be an octave. Whatever may be the rate of rotation of the disc, the frequencies of the two notes must be in the ratio 80 to 40 or 2 to 1, whence we conclude that this ratio, which is called the *Vibration Ratio* of the interval is characteristic of the octave, and the fraction 2/1 may be taken as the measure of the interval of an octave.

If now we make a similar experiment with the 60 circle and the 40 circle, we find the interval is always a fifth, whence we infer that the vibration ratio of a fifth is 3/2.

If the disc is turned at such a rate that the 40 circle gives the middle C of the Pianoforte, the four circles will give the common chord of C. If the disc is turned a little faster they will give the chord of D, and so on.

16. The Measure of Intervals. We may conclude therefore that, if we have any two notes with frequencies m and n, the interval between the notes may be measured by the fraction m/n, for however m and n may vary, the interval will be judged by the ear to remain the same, provided their ratio m/n remains constant.

17. Consonant Intervals. This is true of any interval whatever, but certain intervals are found to have an effect so pleasing to the ear that they are classed as *Consonant Intervals*, and have special names assigned to them. The consonant intervals within the limits of an octave are given below with their vibration ratios. The first five and the last of these ratios can be verified directly with the disc siren as described above.

Octave	$\frac{2}{1}$	Minor Third	$\frac{6}{5}$
Fifth	$\frac{3}{2}$	Major Sixth	$\frac{5}{3}$
Fourth	$\frac{4}{3}$	Minor Sixth	$\frac{8}{5}$
Major Third	$\frac{5}{4}$		

We shall make frequent use of these intervals, and the student should make himself familiar with their vibration ratios.

10   NATURE OF SOUND   [CH. I

18. The Sum of two Intervals. To find the vibration ratio of the interval obtained by adding together two intervals we multiply together the ratios of the two constituent intervals. Suppose we have three notes, which we will call p, q, and r, and suppose p has the lowest, and r the highest pitch of the three. Let us find the vibration ratio of the interval between p and r, when the interval between p and q is a fourth, and the interval between q and r is a major third. The vibration ratio of a fourth is 4 to 3 and that of a major third is 5 to 4. If then the frequency of p is taken as unity, that of q will be 4/3. The frequency of r is 5/4 times as great as that of q, and is therefore $4/3 \times 5/4$ or 5/3. Thus we find that the frequency of r is 5/3 times as great as that of p, or the vibration ratio of the interval p to r is 3 : 5, which corresponds to a major sixth. Similarly a major third and a minor third make $5/4 \times 6/5$ or 3/2, which is the vibration ratio of a fifth. It is evident that the rule holds generally, and we may say therefore that if we add the intervals whose vibration ratios are m/n and m'/n' we get an interval whose ratio is mm'/nn'.

19. The Difference of two Intervals. We can deduce at once from this the converse proposition. If we take the interval whose ratio is m/n from the interval whose ratio is m'/n', the difference is an interval whose ratio is $m'/n' \div m/n$; for by the preceding rule the interval m/n added to the interval $m'/n' \div m/n$ or $m'n/mn'$ gives the interval $m/n \times m'n/mn'$ or m'/n'.

Thus, to add intervals we multiply their vibration ratios together, and to subtract we divide the ratio of the larger interval by that of the smaller.

20. The Diatonic Scale. It will be convenient to give here the vibration ratios that define the intervals between each note of the ordinary Diatonic Musical Scale and the lowest note or *tonic* of the scale. The reasons for the choice of these particular intervals will be given in Chapter XIII.

Since the *intervals* remain the same whatever note is taken as the tonic, it is of no consequence what note we

NATURE OF SOUND

choose as the tonic. We shall take the scale of the white keys of the pianoforte, which have C as their tonic.

C	D	E	F	G	A	B	c
1	$\frac{9}{8}$	$\frac{5}{4}$	$\frac{4}{3}$	$\frac{3}{2}$	$\frac{5}{3}$	$\frac{15}{8}$	2

The meaning of the table is this. Whatever may be the frequency of C, that of c, an octave higher, will be twice as great; that of E will be one and a quarter times as great, and similarly for the other notes. The fractions below the notes are proportional to the frequencies of the notes, in whatever octave on the pianoforte we may take them. The scale can be extended indefinitely upwards and downwards. In order to rise an octave we double all the frequencies, and to fall an octave we halve them.

It will be found by reference to the table in Par. 17 that the interval between

C and c	is an	Octave
C and G	is a	Fifth
C and F	,,	Fourth
G and B or C and E	,,	Major Third
E and G or A and c	,,	Minor Third
C and A	,,	Major Sixth
E and c	,,	Minor Sixth

21. Pitch Notation. It is often convenient to be able to define the pitch of a note in the extended scale, by using a different notation in each octave. We shall use for this purpose the notation introduced by Helmholtz. The letters of the table just given denote the octave from C below the Bass Clef to c near the middle of the Bass Clef. The octave below this is denoted by capital letters with the suffix 1, as C_1, D_1, etc. The octave next below this has the suffix 2, as C_2, D_2, etc. Going upwards from the c near the middle of the Bass Clef we have first an octave with small letters c, d, e, etc., next an octave with the affix 1 as c^1, d^1, etc., then c^2, d^2, etc., and so on.

Thus the successive C's of the pianoforte scale are C_1, C, c, c^1, c^2, c^3, c^4, c^5.

22. Quality. We have now only the quality of the note to consider. This is less simple than the loudness and pitch, and the experiments shewing the relation between the nature of the vibrations and the quality of the resulting note are too complex to be given at this stage.

We can however make a conjecture as to what feature of wave motion is likely to affect the quality of the corresponding note. Two trains of waves may differ in three, and only three, ways. They may have different *amplitudes*, different *wave-lengths*, and different *shapes*. What is meant by *shape* is most easily shewn by a diagram.

Fig. 4 represents two waves, which we may regard as travelling towards the right. They have the same amplitude and the same wave-length, but different shapes. A is symmetrical on the two sides of the crest, whilst B is steeper in front of the crest than it is behind. It is clear that we may have an infinite variety of shapes whilst keeping the same amplitude and wave-length.

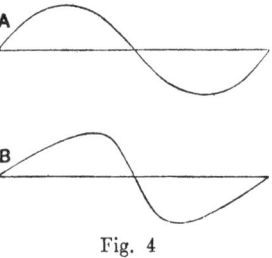

Fig. 4

Now amplitude and wave-length have already been appropriated as characterizing loudness and pitch, and so we have nothing left but shape to characterize quality.

The manner in which quality depends on shape will be considered in Chapter IX.

CHAPTER II

ELASTICITY AND VIBRATIONS

23. Origin of Sound. We saw in the preceding chapter that sound always takes its rise from some vibrating body. It is possible to make a body vibrate by means of a system of cogwheels and levers, without making use of the elasticity of the parts, but vibrations so produced are of little importance for our purpose. Sound almost always arises from vibrations which are due to the elasticity of the vibrating body. Even when the vibrations are not originally elastic vibrations, they become so as soon as they are communicated to the air. Hence it is necessary to give some consideration to the nature of elasticity, and to the vibrations arising from elasticity.

24. Nature and Limits of Elasticity. A body is said to be elastic, if on being deformed in any way by an amount not too great, it tends to return to its original state. Stretch a spiral spring for instance. It tends to return to its original length, and does so as soon as the stretching force is removed. Bend a thin metal rod, and it tends to straighten itself again. Twist the same rod, and it tends to untwist. Close the outlet of a bicycle pump, and press down the handle. The air in the barrel is compressed, but as soon as the pressure is removed, it returns to its original volume.

In most cases of elastic deformation of solids there are limits beyond which the deformation must not go, if the body is to recover its original state. If, for instance, a spiral spring is stretched to a very great extent, it may be that it will acquire a certain amount of permanent stretch, and will not return to its original length. It is then said to have been deformed beyond the *limits of elasticity*.

The limits differ very much for different substances, and for different kinds of deformation. A spiral spring of large diameter and made of steel of good quality might be stretched to double its length without acquiring a permanent set. A straight wire of the same length and material as the spring could be stretched only a very little without suffering permanent deformation, or breaking; whilst a similar wire made of lead would have extremely narrow limits of elasticity, either for bending or stretching. Nevertheless, lead is elastic within its narrow limits, for if a long lead pipe or rod is tapped at one end, the sound is carried along the material to the other end, and we shall see later that sound invariably travels through a body by virtue of the elasticity of the body.

25. Imperfectly elastic solids and viscous liquids. Some substances are elastic if the deforming force acts only for a short time, but acquire a permanent set if the force is maintained. Take, for instance, a rod of pitch or sealing wax, and fix it horizontally by clamping one end. If a small weight is hung on the other end the rod will be bent down a little, and if the weight is soon removed, the rod will recover its original position. If however the weight be allowed to remain, the end of the rod will gradually sink down, and will not return when the weight is removed. It is not, in fact, necessary to put any weight on the end, for the weight of the rod itself will cause it to sink gradually. A lump of pitch placed on a flat horizontal surface will in the course of a few weeks spread out into a thin cake. It behaves in this case as a very viscous fluid, whilst, if the deforming force acts for a very short time, as it does in the case of sound vibrations, the substance behaves as an elastic solid.

There are many substances that behave like pitch, and it has even been suggested that no substance is really solid, but that every body gives way gradually to forces however small, if long enough maintained. This is mere speculation, for the great majority of the bodies which we regard as solid shew no signs whatever of such gradual deformation, even though the forces have lasted for centuries. There is no indication that the Pyramids are flattening under the action of gravity, nor that metals shew any such effect, for ancient coins still shew their inscriptions sharply marked.

A distinction must be drawn between very viscous liquids such as pitch, and solids with narrow elastic limits such as lead. For forces acting for a short time, pitch has wider elastic limits than lead, yet lead is a real solid. There is no reason to suppose that very small forces give it a permanent deformation, however long they act on it.

26. Elasticity of Liquids and Gases. Liquids and gases can be deformed in only one way, namely by alteration of their volume. They offer in general no *permanent* resistance to changes of shape, though, as we have seen, a viscous liquid may offer resistance to rapid changes of shape. They cannot be said to have any limits of elasticity, for, however great the pressure applied to a liquid or gas, the volume will return to its original value when the pressure is removed, provided the other original conditions, such as the temperature, remain the same, or are regained when the pressure is removed.

This statement is not strictly true of liquids, for it has been shewn that in exceptional cases a liquid can be made to break under tension. The exception has no practical bearing on Acoustics.

27. Relation between the Deformation and the Force which causes it. We must next consider the relation between the magnitude of the deforming force and the amount of deformation. For the experimental determination of this relation a spiral spring is convenient, as its limits of elasticity are wide.

Hang up the spring by one end, and at the other end fix a scale pan, and a pointer with a graduated scale behind it.

If it is found that any of the coils of the spring are in contact with each other, put such a weight in the pan as is sufficient just to

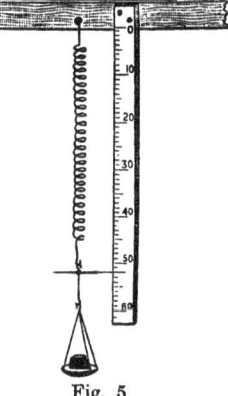

Fig. 5

separate all the coils, and take the reading of the pointer on the scale. Now put such a weight in the pan as will lengthen the spring by an amount that is easily measured. The weight required will depend on the length and stiffness of the spring. Read the new position of the pointer on the scale. Add another similar weight, and read the position again, and so on, repeating the operation several times.

Suppose, for instance, that a weight of 10 gm. lengthens the spring by 1 cm., then the second 10 gm. will be found to lengthen it by another centimetre, and so on, or in other words the lengthening is proportional to the added weight. This proportionality of the deformation to the force is strictly true only for small deformations. In the case of a spiral spring it holds only so long as the coils are approximately horizontal. For a spring a foot long and an inch in diameter, for instance, there would be no great deviation from the law up to an extension of 3 or 4 inches, but if the extension were considerable, say a foot or more, it would be found that the force increased more rapidly than the elongation, even though the elastic limit were not passed. The law holds not only for the extension of a spring, but also for its compression. If a weight of 10 gm. placed in the pan stretches it 1 cm., an upward force equal to the weight of 10 gm. applied to the lower end of the spring will shorten it by 1 cm.

28. Hooke's Law. The law of proportionality of the deformation to the force applied holds for all small distortions of elastic solids, and is known as *Hooke's Law*. The law may be stated thus :—Any small distortion of an elastic body in proportional to the distorting force. As further illustrations we may take the following cases. Fix a rod horizontally by clamping one end in a vice, and hang weights from the free end. The depression of the end will be found to be proportional to the weight applied. Stretch an elastic string or wire horizontally between two points, and hang weights to its middle point; the deflexion will be proportional to the weight. Fix a rod at one end, and apply a couple to the other end, so as to twist it; the angle through which the end is twisted will be proportional to the couple.

Hooke's Law as originally stated applied only to the deformations of solids, but there seems little reason why it should not be taken to include the compression of liquids and gases also, for in the case of these too the change of pressure is proportional to the change of volume, when this change of volume is very small.

It is easily seen that when a gas is *greatly* compressed the added pressure is not proportional to the diminution of volume, for Boyle's Law states that for a gas at constant temperature the product of pressure and volume is constant.

Suppose a column of gas is enclosed by a piston in a cylinder a foot long, and is at the ordinary atmospheric pressure of 15 lbs. per sq. in. Press down the piston 3 in. The gas is now reduced to three quarters of its original volume, and its pressure is consequently $15 \times 4/3$, or 20 lbs. per sq. in. Press down the piston another 3 in., and the pressure becomes 2×15, or 30 lbs. per sq. in. Thus the first 3 in. requires an added pressure 5, whilst the second 3 in. requires 10, or the pressure increases more quickly than the diminution of volume. The law of proportionality of added pressure to diminution of volume can in the case of gases be assumed only for very small compressions. We shall see later that the divergence from the law, when the compression is not very small, gives rise to Combination Tones.

The law of proportionality may be taken as holding generally for liquids, for in their case the resistance to compression is so great that the compression is always small.

29. Forces of Restitution. If an elastic body is deformed, as, for instance, when a rod clamped at one end has the free end drawn aside from its position of rest, the deformation gives rise to internal forces, which tend to bring it back. These are called *forces of restitution*. It is evident that if the rod is held at rest with any given amount of deflexion, the force required to cause this deflexion will be balanced by an equal and opposite force due to the elasticity of the rod. Hence the force exerted by the rod is proportional to the displacement, and is in the opposite direction to the displacement.

30. Potential energy of a Deformed Body. Work has to be done on the rod to displace it, and the rod acquires potential energy equal in amount to this work. Work done by a constant force is measured by the force multiplied by the distance through which it acts. In the case of an elastic body the force increases with the deformation, and hence we must take the average force, and multiply by the total deformation to get a measure of the work. Where the force increases proportionally to the deformation, the average force is the force for half the final deformation.

Hence if a is the displacement, the work is $\frac{1}{2}ka \times a$ or $\frac{1}{2}ka^2$, where k is the coefficient of elasticity for the particular kind of deformation we are considering, that is, the force which will give the unit displacement.

We see then that the potential energy of an elastic body, which has been deformed by forces appropriate to the kind of deformation in question, is proportional to the square of the displacement.

31. Vibrations due to Elasticity. Now release the body, and the forces due to its elasticity at once begin to draw it back to its equilibrium position with increasing velocity, thus transforming the potential energy into kinetic. When the body reaches its equilibrium position, there is momentarily no deformation and no elastic force, and the potential energy has been entirely converted into kinetic. The momentum the body now possesses causes it to pass through its equilibrium position, and to swing out to a distance a on the other side; when its kinetic energy will again have been converted into potential. In this position it has no kinetic energy, but is momentarily at rest, and so falls back, and if there were no dissipation of energy, it would continue to vibrate between the limits $+a$ and $-a$. There will however in general be losses of energy. Part of the energy of vibration will be transferred to the air in the form of waves of compression and rarefaction, as we shall see later; part will be spent in warming the body itself in consequence of internal viscosity, which acts similarly to friction; and there may be other causes of loss of energy, such as skin-friction, or if the vibrating body happens to be magnetised, there may be production of electrical currents in neighbouring bodies. The result of these losses is that the vibrations gradually die down in such a way that the amplitude of any one elongation is in a constant ratio to the amplitude of the next.

32. Isochronism. Hooke's Law leads to the important result that the vibrations of a body due to its elasticity are *isochronous*; that is, the time of performing one complete oscillation is the same whatever the extent of the oscillation.

A simple instance of a body that performs isochronous vibrations is the pendulum. The force that acts on the pendulum is gravity, and not an elastic force, but if the arc of vibration is small, the relation between the force and the displacement is the same as for elastic forces and the vibrations are consequently isochronous. A familiar instance, where the vibrations are due to elasticity, is the balance wheel of a watch. Here the vibrations are due to the elasticity of the hair spring, and are maintained by the main spring acting through the train of wheels and the escapement. If for any reason such as increased friction, or diminished force in the main spring through the watch being nearly run down, the arc of vibration of the balance becomes smaller, the time of vibration is not appreciably altered. The balance continues to vibrate at the same rate whatever its amplitude, and the rest of the watch is merely a contrivance for maintaining the vibrations and counting and recording their number.

33. Proof of the Isochronism of Elastic Vibrations. It is easy to see in a general way why elastic vibrations are isochronous.

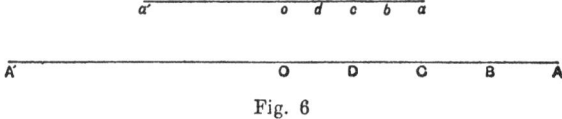

Fig. 6

Suppose we have two particles, one of which can vibrate about an equilibrium position *o*, and the other about an equilibrium position *O*. Suppose also that the two particles have the same mass, and that the same elastic forces act on them when they are at the same distances from o and O respectively.

Draw the first particle aside to a and release it. It will vibrate between the limits a and a', where $oa = oa'$. Draw the second particle aside to A, where $OA = 2oa$. It will vibrate between A and A'. The vibrations of the two particles may then be taken to represent two vibrations of the same particle, one of the vibrations having double the amplitude of the other.

Divide oa and OA each into the same number of equal parts. If the number of parts is very large, each of the parts will be very small, and the force of restitution may be taken as constant over any one part, and equal to its average value over that part. Let f be the average value of the force over the part ab. Then if m is the mass of the particle, and a its acceleration, we know that $f = ma$ or, if the force is proportional to the displacement, the acceleration is so also. Also, since we are assuming f is constant over the part ab, we may assume that a is constant.

Now it is known that if a body starts from rest with constant acceleration a, and moves for a time t, it will pass over a space s given by the equation $s = \frac{1}{2}at^2$, or if $ab = s$, the particle will reach b after a time $\sqrt{\dfrac{2s}{a}}$, and when it reaches b it will have a velocity at or $\sqrt{2as}$.

Now consider the particle that was drawn out to A before being released. Since $OA = 2oa$ the acceleration will now be $2a$ and AB will be $2oa$ or $2s$. Hence the particle will reach B in a time $\sqrt{\dfrac{2 \cdot 2s}{2a}}$ or $\sqrt{\dfrac{2s}{a}}$, the same as in the case of the particle drawn out to A. Its velocity on reaching B will be $2at$ or $2\sqrt{2as}$.

Thus it follows that the two particles reach the end of their first stage in the same time, but the particle with the greater amplitude arrives with double the velocity.

Next consider the second stage bc. Let V be the velocity with which the first particle reaches b and β its average acceleration over bc. Then the second particle reaches B with a velocity $2V$, and during the stage BC it has an acceleration 2β, since the centre of CB is twice as far from O as is the centre of cb from o.

We know that if a body has an initial velocity V, and moves over a space s with uniform acceleration β in time t, then $s = Vt + \frac{1}{2}\beta t^2$. From this equation we can find the time the first particle takes to pass over the stage bc.

If we form a similar equation for the second particle, we have to replace s by $2s$, V by $2V$ and β by 2β, which makes no change in the equation, as we have merely multiplied both sides by 2. Thus each of the particles will take the same time in traversing the second stage, and the process can evidently be continued until they reach o and O respectively. OA has been taken as being twice oa merely for the sake of simplifying the equations. The result would be the same whatever multiple OA is of oa; the particles would reach their equilibrium positions in the same time whatever their amplitudes. It is obvious from symmetry that the time the particle takes in moving from A to O is one quarter of the time it takes in going from A to A' and back again to A, and therefore the time of a complete vibration is the same whatever the amplitude, or the vibrations are isochronous.

34. Simple Harmonic Vibrations. Such vibrations are called *Simple Harmonic Vibrations*, and are of great importance in the theory of Sound, as a vibration of this kind, and of this kind only, gives the sensation of a pure tone of definite pitch with no admixture of tones of other pitches. Any other vibration than a Simple Harmonic Vibration gives rise to a note, which can by suitable appliances be resolved into two or more tones of different pitch. The note arising from a Simple Harmonic Vibration cannot be so resolved. We shall return to this point later.

If elastic vibrations had not been isochronous, music in its present form would have been impossible. Suppose, for instance, that the law connecting amplitude and number of vibrations per second had been that the one was proportional to the other. We have seen that pitch is determined by the number of vibrations per second, and loudness by the amplitude. We should therefore have the result that the louder a note, the higher its pitch. It would be impossible to keep the instruments in the orchestra in tune with each other, and a crescendo would mean a rise in pitch of the whole orchestra.

35. Geometrical Illustration of Simple Harmonic Vibration. The nature of a Simple Harmonic Vibration can be shewn by the following useful geometrical method.

Suppose a point P moves with uniform speed round the circumference of a circle of radius a.

Drop a perpendicular from P on any diameter AA'; then we can shew that N, the foot of the perpendicular, describes simple harmonic vibrations in the line AA'.

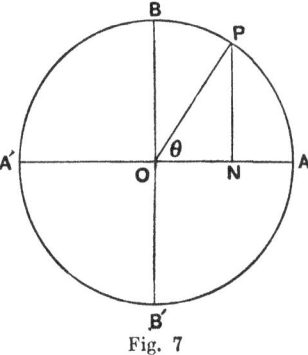

Fig. 7

The radius OP revolves with uniform angular velocity, which we will denote by ω. The acceleration of P is therefore $a\omega^2$ in the direction PO, and the acceleration of N is the component of this in the direction AO, or $a\omega^2 \cos \theta$, where θ is the angle POA. Since $ON = a \cos \theta$, we can write the acceleration of N in the form $\omega^2 ON$, or the acceleration of N is proportional to its distance from O. If N is a particle of mass m vibrating in the line AA' in consequence of a force directed towards O, the force required to give this acceleration must also be proportional to ON, since $F = ma$. This relation of force to displacement is in accordance with Hooke's Law, and N describes simple harmonic vibrations.

The period of vibration of N is the time taken by the radius OP to make one complete turn, or $2\pi/\omega$.

If F is the elastic force when the displacement is unity, the force for displacement ON is $F \times ON$, and the acceleration is $\dfrac{F}{m} \times ON$. The geometrical method gives $\omega^2 \times ON$ as the

ELASTICITY AND VIBRATIONS

acceleration. We can therefore use Fig. 7 to represent the vibrations of a particle of mass m vibrating with amplitude a under the action of an elastic force of magnitude F for unit displacement, if we describe a circle of radius a, and make the radius OP revolve with such an angular velocity that $\omega^2 = \dfrac{F}{m}$; and further, since $\tau = \dfrac{2\pi}{\omega}$, we see that the period of vibration of the particle will be $2\pi \sqrt{\dfrac{m}{F}}$.

This expression is independent of the radius. It is the same whatever the radius of the circle, provided ω has the value $\sqrt{\dfrac{F}{m}}$. Hence this method of treating simple harmonic vibrations leads also to the conclusion that their period is independent of their amplitude, or they are isochronous.

36. Method of Calculating the Period of Vibration. If we know F and m in any particular case, we can calculate the period from the expression $\tau = 2\pi \sqrt{\dfrac{m}{F}}$. We must be careful to use a consistent system of units in the calculation. If, for instance, we use the C.G.S. system, m will be the mass of the vibrating body in grammes and F will be the force in dynes required to give a displacement of one centimetre.

We have taken the simplest case, where the particle vibrates in a straight line, and nothing that is moving has any inertia except the particle itself. We cannot secure this exactly in practice, but a mass suspended by a light spiral spring approximates to it. Suppose the mass of the pan and the body placed in it is M gm., and suppose an additional mass m gm. causes it to sink n cm., then m/n gm. would depress it 1 cm. The force with which m/n gm. is attracted to the earth is mg/n dynes, where g is 981 cm. per second per second.

Consequently the period of vibration of the mass is

$$2\pi \sqrt{\dfrac{Mn}{mg}}.$$

If the period calculated in this way is compared with the period observed directly, it will be found to be a little too small, as we have underestimated M. The spring is moving, and adds something to the inertia. The lowest part of the spring moves as much as the suspended body and the highest part does not move at all. It is plain therefore that we ought to add something less than the mass of the whole spring to the mass of the body. It can be shewn that one third the mass of the spring should be added.

The force required to give unit deformation of any kind to any elastic body is called the *coefficient of elasticity* for the particular body and the particular kind of displacement.

The expression we have found for τ will give the period of elastic vibrations of any kind, if F and M are suitably expressed. F will not always be a simple force, and M will not always be a mass. Suppose, for instance, a body is hung by a wire, and is turned round so as to twist the wire a little. When it is released, it will perform rotational vibrations; the wire twisting first in one direction and then in the other, whilst its axis remains at rest. In this case the coefficient of elasticity F with which we are concerned is the couple that will twist the end of the wire through the unit angle; and the inertia term is the moment of inertia of the suspended body about its axis of rotation.

37. Method of tuning an Elastic Body. In all cases the greater F is, or the "stiffer" the body is to displace, the less τ will be; and the greater the inertia, the greater τ will be.

It is useful to remember this when we have to tune a vibrating body. A tuning-fork gives a good instance. In this case the bending which gives rise to the elastic forces is chiefly at the base of the prongs, and the motion is chiefly at the free ends. If then we scrape or file the prongs near the base, we shall diminish F without making much change in M, and so shall lower the pitch. If on the other hand we file the prongs near the free ends, we shall diminish the inertia without altering the elasticity much, and shall raise the pitch. This is the usual method of tuning a fork. If we wish to lower the pitch of a fork temporarily, we can do it by sticking a little wax on the ends of the prongs, and so increasing the inertia.

38. Period. Amplitude. Phase.

The three main characteristics of a simple harmonic vibration are its period, its amplitude and its phase.

We have already defined the *period* as the time occupied by one complete to and fro vibration.

The *amplitude* is one half the extreme range of the vibration, or the distance between the equilibrium position and either of the points at which the vibrating particle is momentarily at rest. According to this definition the amplitude of N in Fig. 7 is OA or a.

The *phase* of the vibration at any moment is the state of the vibrating particle as regards its position and its direction of motion at that moment. Whenever, for instance, the foot of the perpendicular in Fig. 7 is passing through a particular point, and is moving say from right to left, it is in the same phase. The radius OP rotates uniformly in the same direction, and the position and direction of motion of N is known, if the position of OP is known, and hence the phase can be measured by the angle θ. The term phase is most commonly used in speaking of the difference of phase between two points vibrating with the same period. If they are imagined both to be vibrating in the line AA', though not necessarily with the same amplitude, each will have its rotating radius; and since the particles have the same period, the rotating radii must complete one circuit in the same time making a constant angle with each other, and this angle measures the difference of phase.

The fraction this angle is of the whole circuit is the fraction of a period that one particle is behind the other. If for instance the angle is 90°, we may say that one particle is a quarter of a period behind the other, and this is the most usual way of expressing difference of phase. If expressed as an angle the difference of phase would be said to be $\pi/2$.

If the particles pass through their equilibrium position at the same moment but in opposite directions, they differ in phase by half a period, or, as it is often expressed, they are in opposite phase.

39. The Sine Curve.

The position of the particle N at any time can be shewn by a curve as follows. Divide the circumference of the circle into any number of equal parts beginning at B, and going round the circle in the direction $BAB'A'$. Take a straight line of any length and divide it into the same number of equal parts.

As P moves with uniform speed, each of the sections into which the circumference is divided will be traversed in the same time, and the points marking the ends of the divisions may be regarded as marking a time scale. If a perpendicular be drawn to AA' from the end of each division, and the distance from O of the foot of the perpendicular be measured, we shall have the displacement of N at the ends of a series of equal intervals of time. Now transfer the displacement corresponding to each dividing point on the circle to the corresponding dividing point on the straight line. If N is to the right of O, draw an ordinate equal in length to this displacement upwards from the corresponding dividing point on the straight line. If N is to the left of O, draw the ordinate downwards. Draw a smooth curve through the ends of the ordinates and we shall get a curve such as that shewn in Fig. 8.

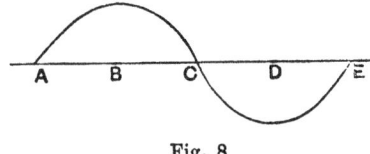

Fig. 8

The curve is drawn only for one complete vibration, but it could evidently be continued indefinitely to the right to represent any number of vibrations. The curve enables us to find the position of the point at any time, for distances measured along the horizontal line from A are proportional to the time elapsed from the moment when the particle was passing through its equilibrium position to the right, and the ordinate at the point corresponding to any given moment shews the displacement at that moment—to the right if the ordinate is above the axis, to the left if it is below.

The curve is known as the *Sine Curve*, for if the maximum ordinate is of unit length, and the length AE is taken to represent 360°, the ordinate at any point in the line will give the sine of the angle corresponding to that point—the sine being positive when the ordinate is above the axis, and negative when it is below.

40. Relation of Velocity to Displacement. The velocity of the vibrating particle when passing through any point of its path can also be shewn by a sine curve.

Referring to Fig. 7, it will be seen that, at the moment for which the figure is drawn, the velocity of N is the component of the velocity of P in the direction NO, or $v \sin \theta$. Hence, since v is constant, the velocity of N is proportional to $\sin \theta$ and so can be shewn by a sine curve. The curve will be displaced a quarter of a period to the left as compared with the displacement curve, for the velocity is a maximum when the displacement is zero, and vice versa. The two curves are shewn in Fig. 9, where points in the same vertical line correspond to the same moment. In the lower curve ordinates above the axis denote velocities to the right, and ordinates below the axis denote velocities to the left.

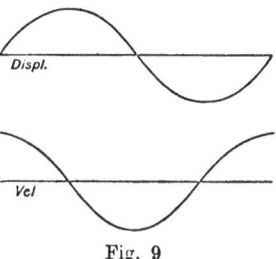

Fig. 9

41. Composition of Simple Harmonic Motion with uniform motion in a straight line. We shall next discuss the curve traced out by a particle which executes harmonic vibrations, and has at the same time some other motion impressed on it.

The simplest instance is the case where the particle executes its harmonic vibrations in one line, and is at the same time carried with uniform velocity in a direction at right angles to that of the vibrations. We can realize this case and trace the required curve in the following way.

Fix a bristle or wire to the prong of a tuning-fork by a small piece of wax, and, whilst the fork is vibrating, hold the end of the bristle against a sheet of paper, which has been smoked with burning turpentine or camphor. The bristle will trace on the paper a straight line, whose length shews the range of vibration of the fork. Now draw the fork uniformly in the direction of its own length, and instead of a straight line we shall get a sinuous curve on the paper.

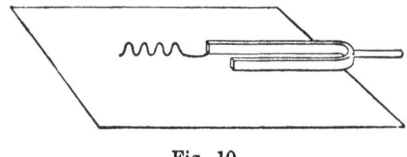

Fig. 10

From what has been said before it is clear that, if the fork is moved with uniform velocity, the curve will be a sine curve.

The straight line that would have been traced by the fork, if the prongs had not been vibrating, is the axis of the curve. Equal distances along the axis correspond to equal intervals of time, and the ordinate drawn from any point of the axis to the curve gives the displacement of the prong at the moment determined by the position of the foot of the ordinate on the axis.

This method of converting a simple harmonic vibration into a sine curve forms the basis of one of the most accurate methods of finding the frequency of a tuning-fork, as will be seen later.

42. Composition of two Simple Harmonic Vibrations at right angles. Let us next find what curve we shall get by compounding two harmonic vibrations in lines at right angles to each other.

We can construct a curve that is approximately compounded of two simple harmonic vibrations by means of the apparatus shewn in Fig. 11 and known as the Harmonograph.

A and B are two pendulums suspended so as to swing in planes at right angles to each other. Each pendulum is continued for a few inches above its point of support, and at the top of each a light horizontal lever is attached by a flexible joint. These levers C and D are joined together at E

by a flexible joint, and a pen is attached to the joint. If the pendulum A is alone set swinging in the plane of the paper,

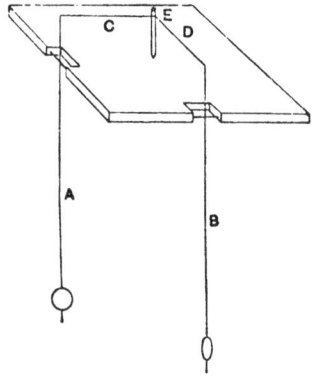

Fig. 11

the joint E will execute approximately harmonic vibrations from right to left. They will not be strictly harmonic for the top of the pendulum B is at rest and the joint E must therefore describe small arcs of a circle round it, but if the levers C and D are 10 or 12 inches long and the arc of vibration of the pendulum is small, the motion of E is nearly enough harmonic for our purpose.

If B is set swinging in a plane at right angles to the plane of the paper and A is at rest, the pen will vibrate harmonically in a direction at right angles to its former direction.

If now both pendulums are set swinging, the motion of the pen will be compounded of the two motions, and a curve will be traced out on the table. If we give an extended meaning to the word "sum," we may say that the displacement of the pen at any moment is the sum of the displacements due to the two separate vibrations at that moment.

There is only one sense in which we can "add" two displacements not in the same straight line, and that is in accordance with the Parallelogram Law used in compounding two forces.

If a particle, when vibrating in the line OP, would have at some moment a displacement OP, and a vibration in the direction OQ would give it at the same moment a displacement OQ, then its actual displacement at that moment must be OR, if each of the components has its full effect. We may imagine the

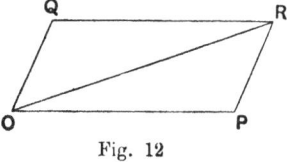

Fig. 12

particle to vibrate in the line OP, and at the same time the line OP to vibrate in the direction OQ always remaining parallel to itself.

The Parallelogram Law, with its extension the Polygon Law, holds not only for the displacements but also for the velocities of vibrating particles.

43. Composition of two Simple Harmonic Vibrations of equal periods. Let us return now to the Harmonograph. The pendulums are generally made with bobs that can be clamped at any point of the rods, so that the times of vibration can be varied. Adjust the bobs to such positions that the time of swing is the same for both pendulums. Now it is clear that E will describe some kind of oval or circular curve. We can shew that a circle is a possible curve.

If P travels round the circle in Fig. 13 with uniform speed, and PN, PN' are the perpendiculars from P on two diameters at right angles, then N and N' describe Simple Harmonic Vibrations in their respective diameters. Their displacements at the moment for which the figure is drawn are ON and ON' and OP is the

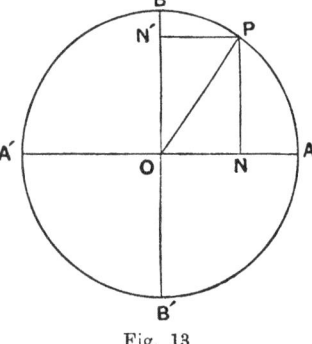

Fig. 13

displacement got by compounding ON and ON' by the Parallelogram Law. This evidently holds for every position of P, and therefore the motion of P may be regarded as obtained by compounding the vibrations of N and N'. These vibrations have the same amplitude and period, but differ in phase by a quarter of a period. Whenever we have these relations between two perpendicular simple harmonic vibrations their resultant is uniform motion in a circle.

If the amplitudes are not the same, it can be shewn that the curve is in general an ellipse, the direction of whose axes depends on the relation between the phases of the constituents. In the particular cases in which the phases are either the same, or differ by half a period, the ellipse degenerates into a straight line.

If AA' and BB' represent the directions and amplitudes of vibration of the two constituents, the curves in Fig. 14 shew five forms of path of the pencil. In No. 1 the two constituents may be said to be in the same phase. When the pencil is passing through O towards the right under the influence of one

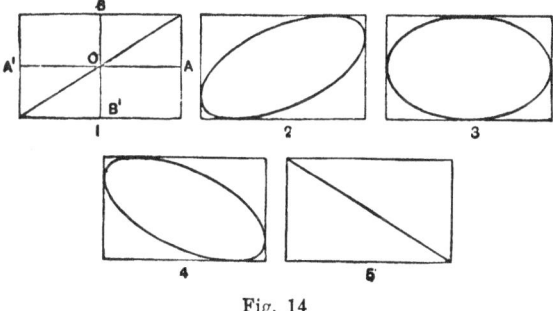

Fig. 14

vibration, it is passing upwards through O under the influence of the other. In No. 2 the difference in phase is between zero and a quarter period. No. 3 shews the case where the difference of phase is a quarter period. The ellipse is then symmetrical about the lines of vibration of both constituents. In No. 4 the difference of phase is between a quarter and a half period, and in No. 5 the two vibrations differ in phase by half a period.

44. Composition of two Simple Harmonic Vibrations of nearly equal periods. If the periods are very nearly, but not quite equal, the curve described during a single complete period of the vibration will be approximately an ellipse; but one of the vibrations will slowly gain on the other, and the difference in phase will change slowly. The curve will then pass through the series of forms in Fig. 14. Beginning say with No. 1, the line will slowly open out into an ellipse, the curve will pass through the forms 2, 3 and 4 to the straight line 5, and then it will pass through the series in the opposite direction until it reaches No. 1 again. If the amplitudes remain constant the curve will always touch the four sides of a rectangle with sides equal and parallel to AA' and BB'.

When the curve has passed from 1 through 2, 3 and 4 to 5, and back again to 1, the phases have returned to their original agreement. Consequently one of the pendulums must have gained exactly one vibration on the other. This gives us a means of comparing their times of swing, for if we find, for instance, that the curve changes from 1 to 5 and back again to 1 in one minute, we know that one pendulum makes one complete vibration per minute more than the other, or the frequency of one is one-sixtieth of a vibration greater than that of the other. This method has been much used in comparing the frequencies of tuning-forks, as will be seen later.

45. Lissajous' Figures. The curves of Fig. 14, and those obtained when the frequencies of the constituents have other aliquot ratios than 1:1, are known as *Lissajous' Figures*.

We will take as our second illustration the case where the frequencies are in the ratio 2:1. We then get in general a figure of 8, degenerating into a parabola for certain relations of phase of the constituents. The relationship between the phases cannot in this case be expressed shortly, as is the case when the ratio of the frequencies is 1:1, for even though the ratio is *exactly* 2:1 the difference of phase is not constant. A vibration in one direction is completed in half the time of a vibration in the direction at right angles to it, and therefore, even though the two vibrations start in the same phase, they immediately come into phases differing from each other. The term difference of phase can in fact hardly be said to be applicable to two vibrations, unless they have the same or very nearly the same period.

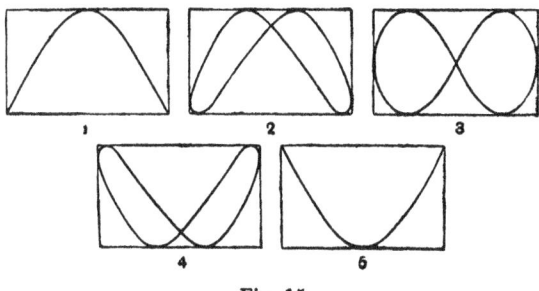

Fig. 15

Fig. 15 shews five forms of Lissajous' Figures for the ratio of frequencies 2:1. The parabolic form of the first and fifth arises when the phases are such that the vibrating point comes to the end of its swing in each of the two perpendicular directions at the same moment. If the ratio is not exactly 2:1 the curve will change gradually back and forwards through the series of figures of Fig. 15.

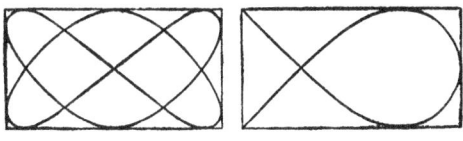

Fig. 16

Fig. 16 shews the symmetrical and the degraded form for the ratio 3:2.

The vibration ratio of the constituents of any of these figures can be found by inspection. In the first curve of

Fig. 16, for instance, the curve touches a horizontal side of the rectangle three times and a vertical side twice. It follows that the tracing pen makes three vibrations in a vertical direction in the time it takes to make two in a horizontal direction, or the ratio is 3 : 2.

When the method is applied to one of the degraded forms, it is to be remembered that the complete path consists of the curve described twice, first in one direction and then in the other. Where the degraded curve comes to a corner of the rectangle, it must be regarded as touching each of the adjacent sides once, and when it has a side of the rectangle as a tangent, it must be regarded as touching that side twice.

46. Optical method of compounding Simple Harmonic Vibrations. Lissajous' Figures can be produced optically by means of two tuning-forks. Each fork has a small mirror on the side of one prong. One fork A has its prongs vertical, and the other B has its prongs horizontal, as shewn in Fig. 17. A beam of light from a small source S

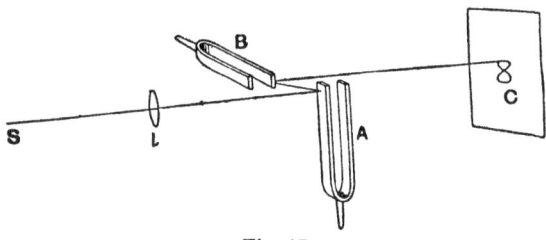

Fig. 17

strikes the mirror on the prong of the vertical fork, is reflected to the prong of the horizontal fork, and thence to a screen C where it is brought to a focus by the lens L. If the fork A alone is vibrating, the spot on the screen describes harmonic vibrations so rapidly that only a bright vertical line of light is seen. If the horizontal fork B alone vibrates, the spot describes a horizontal line. If both forks vibrate, the two vibrations of the spot are compounded, and if the periods of the forks bear some simple ratio to each other, one of Lissajous' figures is seen on the screen.

CHAPTER III

TRANSVERSE WAVES

47. Introductory. In Chapter II we discussed the motion of a single particle executing Simple Harmonic Vibrations. In the present Chapter we shall consider the way in which wave motion arises from the simultaneous vibration of a series of particles. We shall take as simple a case as possible, and the possibility of the motion to be described must be assumed. It can be shewn theoretically to be a possible form of motion of, for instance, consecutive short sections of a stretched string, but the method of proof is beyond our scope. At a later stage in the chapter experiments will be described which are consistent with our conclusions, but cannot be taken as a complete proof of the correctness of the assumptions. This chapter must be regarded rather as a description of the properties of transverse waves than as a logical deduction of those properties from the laws of elasticity.

48. Waves arising from the Harmonic Vibration of a series of particles. Let us suppose we have a series of particles placed at equal distances from each other along a straight line, and each capable of vibrating harmonically in a direction at right angles to that line, the lines of vibration all lying in the same plane.

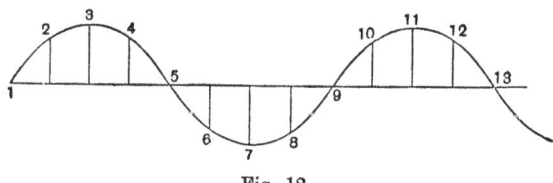

Fig. 18

Imagine the particles to be vibrating with equal periods and amplitudes, and in such phases that each is some constant fraction of its period behind the particle on its left. There is then a constant difference of phase between any two consecutive particles, the retardation of phase increasing as we go from left to right along the series. Suppose, for instance, that each is one eighth of a period behind its left-hand neighbour, and that at the moment we are considering particle 1 is passing downwards through its equilibrium position. Then since 2 is one eighth of a period behind 1, it will not have reached its equilibrium position, but will be moving downwards towards it. Its position can be found graphically from a diagram similar to Fig. 7.

Describe a circle with radius* equal to the amplitude of vibration of each of the particles, and draw a diameter P_3P_7.

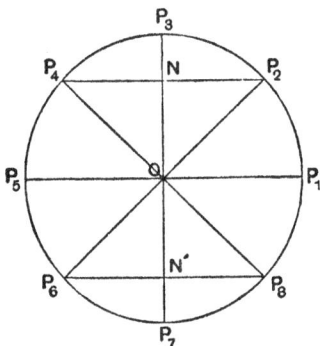

Fig. 19

The motion of any particle of the series is given by the motion of the foot of the perpendicular drawn to the diameter P_3P_7 from the end of a radius which revolves with uniform velocity.

The particle 1 is passing downwards through its equilibrium position and is therefore represented by the foot of the

* (Fig. 19 is drawn on a larger scale than Fig. 18 for the sake of clearness.)

perpendicular from P_1, the radius OP_1 revolving in the direction of the hands of a clock. Particle 2 is an eighth of a period behind 1. Its radius is therefore OP_2, which is an eighth of a complete revolution, or 45°, behind OP_1, and ON is the distance of 2 above its equilibrium position. Particle 3 is 45° behind 2. It is therefore at a distance OP_3 from its equilibrium position, and is momentarily at rest. Particle 4 is at a distance ON above the axis and is moving upwards. Particle 5 is moving upwards through its equilibrium position. Particle 6 is at a distance ON' below, and is moving upwards, and so on.

When the radius reaches P_1 again, we begin a second round of the circle; P_9 is at P_1, P_{10} at P_2, etc.

All the particles between 3 and 7 are moving upwards, and all those between 7 and 11 are moving downwards. 3, 7 and 11 are at the ends of their swings and momentarily at rest.

Thus it appears that at the moment we are considering the particles will lie along a curve such as is shewn in Fig. 18, and it is evident that this is the sine curve, for the method by which we have drawn it is the same as that by which we drew the sine curve.

49. Velocity of the Waves. Next let us consider where the particles will be at a moment one eighth of a period later. 1 will have moved downwards below its equilibrium position. 2 will have reached its equilibrium position and be moving downwards through it. 3 will have moved downwards from the end of its swing and will be as much above the axis as 2 was one eighth of a period earlier; 4 will have reached the end of its swing, and so on.

Each particle will have come into a position corresponding to that occupied by its left-hand neighbour one eighth of a period earlier. Particles 4 and 12 will now be on the crests of the waves instead of 3 and 11, and 8 will be at the bottom of the trough instead of 7. In fact the whole curve has moved to the right by an amount equal to the distance between two particles.

Thus we see that, when a series of particles execute similar simple harmonic vibrations with constant difference of phase as we pass along the series, the result is a wave in the form of a sine curve travelling in the direction in which we find a gradual *retardation* of phase as we pass along the particles.

If in passing along the particles in the order 1, 2, 3, 4, etc. we had found a gradually *advancing* phase, the result would have been similar. The particles would lie as before along a sine curve, but the waves would travel towards the left.

Let us next find with what velocity the wave travels. We saw that in one eighth of a period the crest moves from 3 to 4. In what time will it travel over one whole wave-length? We take as a *wave-length* the distance between any particle and the nearest particle that is in a corresponding position. 1 and 9 for instance are a wave-length apart, as are also 3 and 11. We must not take 1 and 5, for, though they are both passing through their equilibrium position, 1 is moving downwards and 5 upwards.

Let us choose in particular 3 and 11. It is evident from what we said above that the crest which is now occupied by 3 will be occupied by 11 after one whole period, and the wave will travel one wave-length in the time taken by any one particle to execute one complete vibration.

50. Relation between Wave-Length and Velocity.

Now the velocity of the wave is the distance it travels in one second. Let τ be the period of vibration and λ the wave-length, then the number of vibrations a particle makes in one second is $1/\tau$, which we denote by n, and the distance the wave travels in one second is $n\lambda$, whence we have the formula

$$v = n\lambda.$$

The quantity n is what we have called the frequency, and denotes either the number of vibrations executed by any one particle in a second, or the number of waves passing a given point in a second.

The formula $v = n\lambda$ is frequently used in acoustical calculations, as it gives the relation between three important characteristics of a wave train. It is sometimes more convenient to use the equation in the form $\lambda = v\tau$.

51. Non-Harmonic Waves. In finding the wave form of Fig. 18 we have supposed that each of the particles is vibrating harmonically. We have done this because as has already been said, the simple harmonic vibration must be regarded as the fundamental form in the theory of Sound. It is not however necessary to limit ourselves to Simple Harmonic Vibrations. We shall get a progressive wave—though not in the form of a sine curve—if the particles all vibrate in the same way, whatever that way may be. The vibrations must be *periodic*, that is, each particle must continue repeating the same succession of movements, and there must be a regular change of phase as we pass along the series. As we cannot represent a non-harmonic vibration as the projection of uniform circular motion, we cannot express the difference of phase in this case by an angle, but we may express the relations of phase by saying that each particle must pass through its equilibrium position some fixed fraction of a period later than its left-hand neighbour.

For the present we shall confine ourselves to Simple Harmonic Vibrations and leave the more complicated forms until we have discussed Fourier's Theorem.

Waves such as have been described are called *Transverse Waves*, because each particle vibrates in a line transverse to the direction in which the wave travels. In Chapter IV we shall describe a form of waves called Longitudinal Waves, where the line in which the particle vibrates is coincident with the direction in which the wave travels.

52. The Medium does not travel with the Waves. It should be noted that in these, as in all other waves, the vibrating particles do not travel with the waves. Every particle of air, or water, or string, or whatever it may be that transmits the waves, travels repeatedly over the same limited path, and what passes on is a *shape* or arrangement of the particles.

When waves are passing over the surface of water, floating bodies do not travel with them. A cork merely describes a small closed curve as the waves pass it, from which we see that it is not the water itself which travels onwards, but the shape of the surface.

40 TRANSVERSE WAVES [CH. III

*53. The Equation for Simple Harmonic Wave Motion. The equation of the Simple Harmonic Wave of Fig. 18 can be obtained in the following way:

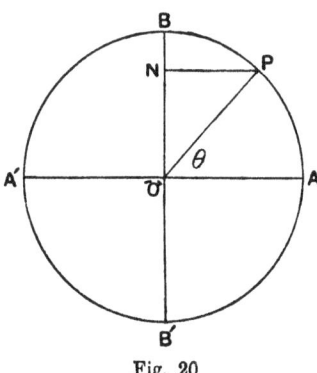

Fig. 20

We shall represent the motion of any particle of the series as before by the motion of the foot of the perpendicular drawn from the end of a uniformly rotating radius on a fixed diameter of a circle. In order to bring the equation into its most usual form we shall measure the angle θ from the diameter AA', but drop the perpendiculars on the diameter BB'.

We have then the displacement $ON = a \sin \theta$, and as θ increases uniformly with the time, we may write $\theta = \omega t$, whence $y = a \sin \omega t$, where y is the displacement of a particular particle at any time t. Every other particle executes exactly similar vibrations, but the phase is uniformly retarded as we pass to the right along the row of particles. That is to say, if x is the distance measured towards the right from the particle whose motion is given by $y = a \sin \omega t$ to some other particle, the angle θ for this other particle will be smaller by an amount proportional to x, say kx, where k is a constant, and therefore this second particle's motion will be given by $y = a \sin (\omega t - kx)$. This equation applies to any particle of the series, including that for which $x = 0$, and is therefore the general equation giving the displacement of any particle at any time, or the equation of the wave.

The period of vibration τ of any particle is the time taken by the radius OP in making one revolution, and is therefore $2\pi/\omega$. It follows that $\omega = 2\pi/\tau$, or, since $\lambda = v\tau$, we have $\omega = 2\pi v/\lambda$.

If we pass from any one particle through a distance λ along the series, we shall arrive at a particle in the same phase as that from which we set out, for everything recurs after a wave-length. It follows that if x changes by λ, θ must change by 2π, or $k\lambda = 2\pi$, from which relation we get $k = 2\pi/\lambda$. Inserting the values we have found for ω and k we find for the equation of a train of harmonic waves

$$y = a \sin \frac{2\pi}{\lambda} (vt - x).$$

54. Transverse Waves on a Stretched String.

We shall next describe a method by which waves such as those which have been described can be shewn experimentally.

Take an indiarubber cord or thick-walled tube, say 15 ft. long. Fix one end to the wall, and holding the other end in the hand, stretch the cord, until it is nearly horizontal. Its weight will cause it to droop somewhat in the middle but this will not affect the results. Suppose B is the fixed end and A is the end that is held in the hand.

Fig. 21

Move the end that is in the hand rapidly a few inches above A, then the same distance below A, and finally bring it back to rest at A. The end has then executed a vibration which approximates more or less closely to a harmonic vibration. One complete wave consisting of a crest and a trough will be seen to run along the cord to the fixed end, where it will be reflected and run back to the hand. At the hand it will be reflected again, and will run back and forwards several times, until the imperfect elasticity of the cord causes it to die away. Assuming from this experiment that a wave *can* travel along such a stretched cord, we can find its velocity.

55. Velocity of waves on a Stretched String.
Let us find what forces act on a short piece PQ in the front half of the wave of Fig. 21, taking the tension in the cord to be T. As the tension is the same everywhere, the element PQ will have a force T acting on each end in the direction of the arrows, and as these two forces act along the tangents at P and Q and are therefore not exactly in the same straight line, they will have a resultant along the line AO bisecting

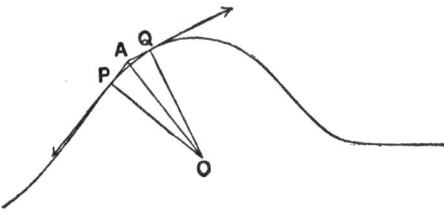

Fig. 22

the angle between them. As PQ is very short, we may assume it is part of a circle. Draw the radii of this circle PO and QO meeting at the centre of curvature, and resolve the two forces along the line AO. If the angle POQ is denoted by θ, the resultant of the forces will be $2T\cos OAP$ or $2T\sin \tfrac{1}{2}\theta$. Since θ is very small we may write $\tfrac{1}{2}\theta$ for $\sin \tfrac{1}{2}\theta$, and the resultant force due to the tension is therefore equal to $T\theta$. Imagine the string to be enclosed in a smooth tube of the shape of the wave, and to be at rest. Then the force $T\theta$ is balanced by the resultant pressure of the tube in the direction OA. Now let the string be drawn through the tube towards the left with a constant velocty v. The small piece PQ is now moving in a curved path, and there is therefore a resultant force on it due to its motion and equal to mv^2/r, where m is the mass of the element and r the radius of curvature of its path at PQ. Hence the force exerted by the tube on the element of the string is now $T\theta - mv^2/r$. Suppose now the velocity is so adjusted that the tube exerts no force on PQ, then $T\theta = mv^2/r$, or $v^2 = Tr\theta/m$. Let ρ be the mass of unit length of the string, then m the mass of PQ is ρPQ, or $\rho r\theta$, and therefore

$$v^2 = \frac{T}{\rho}.$$

The element of string that we are considering now exerts no force on the tube. As the expression for the velocity which secures this result does not contain r, but only T and ρ, which are the same for every element of the string, it follows that there is no pressure on any part of the tube.

The tube is therefore not needed to maintain the shape of the string, and can be imagined to be removed without making any change in the motion. We have then the string running rapidly to the left, and the wave form in equilibrium, and at rest relatively to surrounding objects. That is the same thing as saying that the wave is running along the string. If the string is at rest, and the wave is started running along it, it will travel with a velocity $\sqrt{T/\rho}$, as any other velocity would require the continuous application of external forces to prevent the form and amplitude of the wave changing as it progresses. Thus $\sqrt{T/\rho}$ is the only velocity that is consistent with the wave travelling along unchanged without constraint, as experiment shews that it actually does.

It follows from this expression that the velocity of a transverse wave along a stretched string depends only on the tension and the mass per unit length of the string, and not on the wave length. Waves of all lengths travel with the same velocity, provided that T and ρ remain the same.

56. Wave-length and period of vibration in a stretched string. We saw by consideration of Fig. 18 that $v = \lambda/\tau$. If we substitute the value we have just found for v we find $\tau = \lambda \sqrt{\dfrac{\rho}{T}}$. Thus τ, the period of vibration of a small element of the string, is not independent of the wave-length. It is easy to see that this must be the case, for the shorter the wave for a given amplitude, the greater will be the curvature of the string; and the greater the curvature of the string, the greater will be the resultant of the two forces at P and Q in

Fig. 22. Hence, the restoring force is greater for a short wave than for a long one, and the period of vibration of a particle is less when a short wave is passing over it than it is when a long wave is passing.

57. Superposition of Waves. Let us now return to our stretched cord and make another experiment. Send a wave along the cord, and just as it is reflected from the fixed end, send another along to meet it. They will meet in the middle of the cord and will be seen to pass through each other. Each will go on its way quite undisturbed except at the moment when it is passing the other. At that moment the two waves cannot of course exist separately, for the cord cannot take two different shapes at once. In order to find the actual shape of the cord we proceed in the same way as we did in constructing Lissajous' Figures. The procedure is simpler in this case for, since the displacements are in the same straight line, their resultant is simply their sum or difference, according as they are in the same or opposite directions. We may regard this as a limiting case of the Parallelogram Law, where the angle between the two displacements is zero.

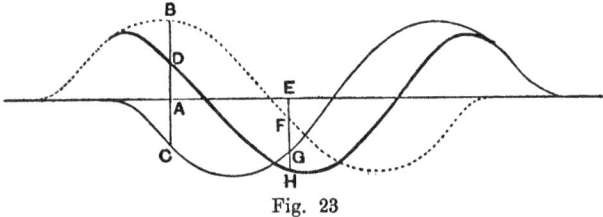

Fig. 23

Suppose the thin continuous curve in Fig. 23 represents the shape the cord would take at a particular moment if only the reflected wave were present, and suppose the dotted curve represents the shape of the cord if the reflected wave were not present, and we had only the direct wave that was sent to meet it.

The shape of the string when both waves are present is found as follows. At any point draw an ordinate at right angles to the undisturbed direction of the string and long enough to cut both curves. Let ordinates measured upwards be considered positive, and those measured downwards negative. Now add the ordinates of the two curves at the point chosen, taking account of their sign, and mark the point that has an ordinate equal to their algebraic sum. At A, for instance, the ordinate of the dotted curve is AB, and that of the full curve is AC. AB is positive and AC is negative. Their algebraic sum is therefore the difference of the two lengths AB and AC, and is positive since AB is greater than AC. Thus we have to mark a point D between A and B, such that AD is the difference between AB and AC, or to reduce the ordinate AB by an amount BD equal to AC. D will be a point on the curve actually assumed by the string at the moment considered. At E both ordinates are downwards, and the ordinate required is EH, such that $EH = EF + EG$. Proceed in the same way at a number of points along the string, and draw a smooth curve through the ends of the ordinates obtained by adding algebraically the ordinates of the two curves to be compounded. The curve so obtained gives the actual shape of the string at the moment we are considering. It is shewn by the heavy curve in Fig. 23.

58. Reflection of Waves at the end of a string.
We must next look more closely into what happens when the wave is reflected at the fixed end of the cord.

Shake the end held in the hand in such a way as to send only a crest along the cord. When it is reflected it will be seen to be changed into a trough, and conversely a trough reflected from the fixed end returns as a crest. If a complete wave consisting of a crest and trough is sent along the cord crest first, it will return trough first.

We may mention here, for the sake of completeness, though the fact has little practical importance for our present purpose, that if the reflection is from the free end of a cord, there is not this change. A crest returns as a crest and a trough as a trough. This can be shewn by letting the cord hang vertically downwards, with the upper end held in the hand and the lower end free. If the hand is moved so as to send a crest down the right-hand side of the cord, for instance, it will return from the free end as a crest still on the right. We shall meet later

46   TRANSVERSE WAVES   [CH. III

with an analogous and important difference between the reflections of an air-wave at the closed and open end of a pipe

It is easy to see that there must be some such difference between the reflection at the fixed and the free end. Suppose B (Fig. 24) is the free end and the crest is just reaching it. B will be pulled to the right by the arriving wave, and as it is free to move, it will acquire a velocity to the right, just as though it had been moved to the right by the hand, and so will send back a crest to the right.

When B is fixed this cannot happen. Whilst the first half of the crest is arriving, B would, if free to move, be moving to the right. But B is fixed, and we may therefore regard it as having also at each instant a velocity equal and opposite to that due to the direct wave. Thus whilst a crest is arriving at B a trough is also leaving it.

Fig. 24

59. Method of finding the position of the reflected wave. We can find the position at any moment of the wave reflected from a fixed end by the following construction.

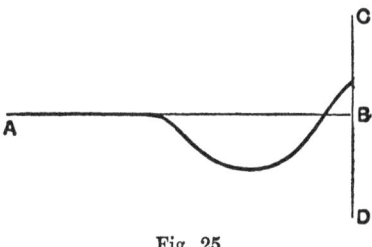

Fig. 25

Let AB be the position of the undisturbed string, the end B being fixed. Suppose one complete wave has been sent from A, and at the moment considered it has reached the position shewn by the full line. This line does not give the shape of the string at the moment in question, for part of the wave has already been reflected, and the reflected part has to be compounded with the arriving part to give the shape of the string.

58–60]   TRANSVERSE WAVES   47

Imagine the arriving part of the wave to be optically reflected twice. First reflect it in AB. We thus get Fig. 25 converted into Fig. 26, No. 1. Next reflect the curve in Fig. 26, No. 1

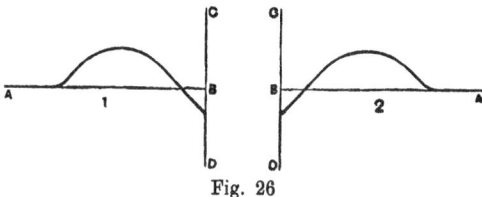

Fig. 26

in CD, and we get Fig. 26, No. 2. The continuation of this curve to the left of CD gives the position of the reflected part of the wave at the moment we are considering, and if we compound it with the arriving part of the wave, we shall get the actual shape of the string at this moment. In order to shew the application of the method we shall pass at once to the important case where a continuous train of waves travels to the fixed end of the string and is reflected there.

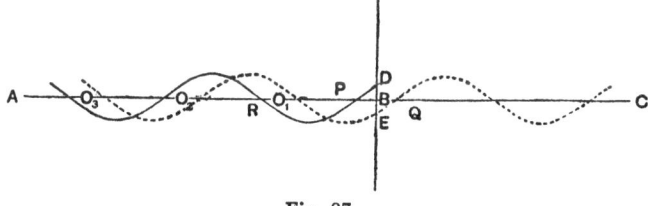

Fig. 27

60. Superposition of the direct and reflected waves. The continuous curve between A and B represents the train of waves travelling towards the fixed end B, where they are reflected. The curve between B and C is the curve obtained by the double reversal of the arriving waves. The dotted curve between A and B is the continuation to the left of the curve BC, and shews the position of the reflected train at the moment for which the figure is drawn. If the full and dotted curves between A and B be compounded by adding their ordinates as described above, we shall get a curve which represents the actual shape of the string at this moment.

At B the ordinates of the direct and reflected curves are equal and opposite, and the ordinate of the resultant curve is consequently zero at this point. This must obviously be the case, as the point B is fixed and cannot have any displacement. Now remembering that the direct curve is moving to the right and the reflected curve with its imaginary extension to the right of B is moving to the left, it is clear that BD and BE will always be equal and opposite. At the moment for which the figure is drawn D and E are both moving towards B.

When P reaches B from the left Q will reach B from the right and both constituent curves will have zero ordinate. A little later D will be below B, and E will be above, but always by the same amount. Assuming as the result of experiment that a train of waves *is* reflected from the fixed end, and observing that the direct and reflected train must always compound to zero at B, we can, in fact, deduce the position of the reflected train relatively to the direct train, and the construction just given is devised to secure that the conditions shall be satisfied. No other position for the reflected train than that shewn will give the resultant ordinate at B always zero.

61. Stationary Vibrations. It will be seen on examining the figure that B is one of a series of points at which there is never any displacement. These points are marked by circles and are obviously half a wave length apart.

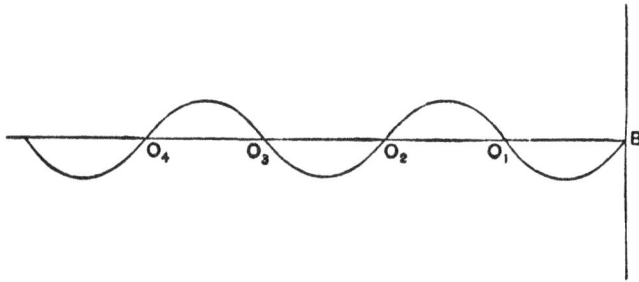

Fig. 28

TRANSVERSE WAVES

If the resultant curve is drawn it will be found to be as shewn in Fig. 28. It is a sine curve of the same wave-length as the original waves, but instead of travelling along it remains in the same position and merely changes in amplitude.

When P reaches B the direct and reflected waves will be coincident, and the resultant curve will have double the amplitude of either constituent. When the direct curve has travelled still further by half a wave-length, and R has reached B, the two constituents will be exactly opposite in phase, and their ordinates will be everywhere equal and opposite. Hence the resultant curve will be simply a straight line. The string will be in its equilibrium position everywhere for the moment.

In the same way a number of other positions of the string can be determined, and it will be found that whilst two consecutive points marked O remain at rest, the string between them takes all the intermediate positions between a crest of double the height of a crest of the direct or reflected wave, and a trough of double the depth of the trough of a direct or reflected wave.

This mode of vibration of the string is called *stationary vibration*. The points that are always at rest are called *nodes* and the portion of string between two consecutive nodes is variously denoted a *loop*, a *ventral segment* or a *vibrating segment*. The point at the centre of a loop is called an *antinode*. It is easily seen from Fig. 27 that the length of a loop is half the wave-length of the original train of waves.

62. Experimental demonstration of Stationary Vibrations. Stationary vibrations can be readily formed on the indiarubber cord. Instead of giving an isolated shake to the end held in the hand as in the former experiment, keep the end in continuous up-and-down vibration, so as to send a train of waves along the cord. At first the result will probably be only a confused motion, but if the rate of vibration be gradually increased, it will be found that at some particular rate the string will break up into a number of loops separated by nodes.

If the rate of vibration be gradually increased still further, confusion will again ensue, to be followed presently by the string again breaking up into loops, but with one loop more than before. We cannot get nodes and loops with *every* rate of vibration of the end, for the wave-length depends on the rate of vibration, and the string will break up into loops only when the half wave-length is an aliquot part of the whole length of the cord.

63. Transmission of energy along a train of Stationary Waves. It is to be noted that when the string breaks up into loops the hand is at a node, though we have spoken of the nodes as points of no motion. The string cannot be absolutely at rest at a node, for if it were, no energy would be transmitted through it to the part beyond. The amplitude of vibration of the hand need be only very small to set up after a time a considerable amplitude of vibration of the string. At the nodes intermediate between the hand and the fixed end the motion is mainly a change of direction of the string, the vibrating string cutting the line given by the undisturbed string at a varying angle. As no string is perfectly flexible this change of direction is sufficient to transmit energy through a node. This will be better understood when the chapter on resonance has been read.

64. Frequency of Vibration of a String. The time taken for a loop to perform one complete vibration, say from the upper limit of its swing to the lower limit and back to the upper limit, is the time taken by the direct or reflected train to travel one wave-length; for the loop is at its upper limit when each of the constituent trains has the centre of a crest at the centre of the loop, and the loop will come again to its upper limit when the next crest of each of the constituent trains has come to the same place.

We saw that the velocity of a train of waves is $\sqrt{T/\rho}$. If l is the distance between two nodes $2l$ is the wave-length, and hence the period of vibration $\tau = 2l \div \sqrt{T/\rho}$ or if we express it, as is more usual, in terms of the frequency n, we have

$$n = \frac{1}{2l}\sqrt{\frac{T}{\rho}}.$$

As there is no appreciable motion at a node, we may now clamp the string at any two nodes, and we get the ordinary case of a string fixed at both ends, as used in musical instruments such as the violin, harp, pianoforte, etc. In the ordinary use of such instruments there are no nodes except those at the two fixed ends, and the number of vibrations per second of a string so used is $\frac{1}{2l}\sqrt{\frac{T}{\rho}}$. In using the formula care must be taken to use a consistent system of units. If the C.G.S. system is used, l will be measured in centimetres, T in dynes, and ρ in grammes per centimetre. If the English system be used, l may be measured in feet, T in poundals, and ρ in pounds per foot.

As an illustration take the case of a string one metre long, weighing 5 grammes per metre, and stretched with a force equal to the weight of 20 kilogrammes. Reducing these to the units mentioned above for the C.G.S. system we have

$$n = \frac{1}{200}\sqrt{\frac{20000 \times 981}{\cdot 05}} = 99.$$

65. Laws of the Vibrations of a String.

The formula $n = \frac{1}{2l}\sqrt{\frac{T}{\rho}}$ shews that we can alter n, the frequency of a stretched string, and consequently the pitch of the note given out by the string, in three ways.

(1) We can alter the length of the string. The frequency is inversely proportional to the length, so that if for instance we halve the length, the string will vibrate twice as rapidly.

(2) We can alter the tension of the string. The frequency is proportional to the square root of the tension, so that if T be made 4 times as great, n will be doubled.

(3) We can load the string by twisting fine wire round it. The frequency is inversely proportional to the square root of the mass of unit length of the string; thus if we quadruple the mass we shall halve the frequency.

66. The Monochord.

These laws can be tested experimentally by means of the monochord. A wire fastened to a pin at one end of a rectangular box passes over two fixed bridges A and B, which may conveniently be a metre apart, and then over a pulley to a scale pan in which weights can be placed.

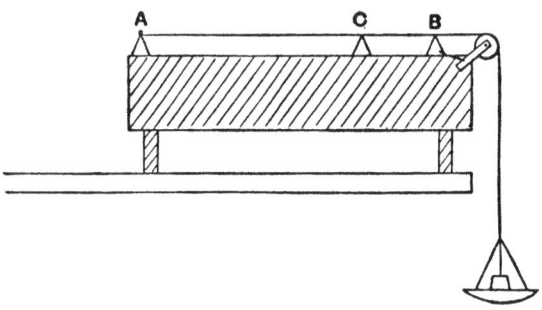

Fig. 29

The tension of the string is then equal to the weight of the scale pan and weights.

A third bridge C can be moved along a graduated scale on the top of the box to any position between A and B.

We have here a string AC, whose length can be altered by moving the bridge C, and whose tension can be altered by altering the weights in the scale pan. We can also substitute a heavier or lighter string.

It is convenient to have stretched on the box a second string for comparison. This is fixed at one end to a pin that can be twisted by means of a key, so that the tension of the string can be altered.

67. To test the relation between the length and frequency of a string. Adjust the tension of the comparison string until on being plucked or bowed with a violin bow it gives a rather higher note than the string AB when the the bridge C is removed. Insert the bridge C and adjust its position until the string AC gives the same note as the comparison string, and read the length AC on the scale. Now adjust C until AC gives a note an octave higher than the comparison string. We know that we have doubled the frequency of AC and by observing the new position of C on the scale we shall find we have halved the vibrating length of the string. Similarly, if we adjust C until AC gives a note a fifth above the note of the comparison string, we shall find we have reduced AC to two thirds the length it had when the strings were in unison, and we know that the rise of pitch of a fifth corresponds to an increase of n in the ratio 3 to 2. Thus in each case $n_1 : n_2 = l_2 : l_1$ where n_1 and l_1 are the frequency and length corresponding to the unison, and n_2 and l_2 those corresponding to the octave or to the fifth.

68. To test the relation between the tension and frequency of a string. Adjust AC to give unison as before. Keeping C fixed, add weights to the scale pan until the note given by AC is an octave above that of the comparison string. It will now be found that the scale pan and weights together weigh four times as much as when the strings were in unison. We thus see that the frequency has been increased proportionally to the square root of the tension. A similar experiment can be performed with the interval of a fifth, when it will be found that the weight must be increased in the ratio of 3^2 to 2^2.

69. To test the relation between the mass of a string and its frequency. Adjust AC to give unison and denote the length AC by l_1. Take off the wire, cut off a piece and weigh and measure it. Suppose the weight per centimetre is ρ_1 grammes. Now fix on the monochord another string of different thickness whose weight per centimetre is ρ_2, and attach the same scale pan and weights. Adjust AC until the new string is in unison with the comparison string, and let the length AC be now l_2.

If the laws be true we have in the first case

$$n_1 = \frac{1}{2l_1} \sqrt{\frac{T}{\rho_1}},$$

and in the second case

$$n_2 = \frac{1}{2l_2} \sqrt{\frac{T}{\rho_2}}.$$

Since the frequencies are the same, $n_1 = n_2$, and we ought therefore to find that $l_1 \sqrt{\rho_1} = l_2 \sqrt{\rho_2}$. The measurements that have been made will be found to satisfy this relation, and consequently to prove the law $n \propto \dfrac{1}{\sqrt{\rho}}$, as we have already verified the law $n_1 : n_2 = l_2 : l_1$, which is also involved in the expression we have just found.

70. Positions of the Nodes. In these experiments with the monochord we have had nodes only at the fixed ends. The experiments with the indiarubber cord shewed that a string can vibrate in any number of equal sections separated from each other by nodes, and it is possible to make the string of the monochord break up into sections, if, when it is plucked, it is touched gently with the finger at some point that is a possible node, so as to prevent vibration at that point.

Remove the bridge C and pluck the string. Now touch the string gently at the middle, and pluck again. The note will be found to have gone up an octave, and the string will continue to give this higher note for a short time after the finger is removed. In the first case there were nodes only at the fixed ends. In the second case the finger compelled the formation of a node at the centre, and consequently the note was that given by a string of half the length.

Next touch the string one-third of the way from one end and pluck the shorter section. The string will vibrate in three sections, and give a note with three times the frequency of that given by the whole string, that is, a twelfth higher. Similarly we can make it vibrate in four, five, etc. sections by touching it one fourth, one fifth, etc. from the end. The notes are produced more easily—especially the higher ones—if the string is touched with something narrower than the finger, such as a match or a quill.

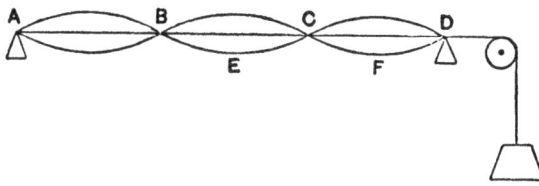

Fig. 30

When the string is touched at B and plucked between A and B, where AB is one third of AD, there is a node also at C, half way between B and D. This can be shewn by cutting short strips of paper, bending them in the middle, and hanging them over the string at E, C and F. When the string is plucked, the paper riders at E and F will jump off, whilst that at C will remain at rest, thus shewing that a node has been formed at C.

We see then that a string can give out a series of notes according as it vibrates as a whole or in two, three, four, etc. sections. If the frequency of the lowest note is taken as unity, the frequencies of the higher notes will be 2, 3, 4, etc., since the frequencies are inversely proportional to the lengths of one section of the string in each case, and the string is divided into two, three, four, etc. equal sections for the respective notes.

71. Nomenclature of Modes of Vibration. Most sounding bodies are capable of vibrating in a variety of ways, which give rise to notes of different pitches. Of these notes the lowest is called the *fundamental* and the rest *overtones*.

A series of notes in which the frequencies are in the ratio 1, 2, 3 etc. is called a *harmonic series*, the separate notes being referred to as *harmonics*. When, as in the case of a string, the overtones form a harmonic series, they may be called *harmonic overtones*. We shall meet with many cases in which the overtones are not harmonic.

72. The Harmonic Series. We shall make frequent use of this series in the following chapters, and the student who has some knowledge of musical notation will find it useful to learn the positions of the lower members of the series on the usual clefs.

In Fig. 31 the first twelve terms are shewn in the positions they would occupy if the lowest note were C. The figures to the right shew the relative frequencies. The seventh and the eleventh harmonics are

Fig. 31

enclosed in brackets as they are not exactly represented by any note in the scale. The seventh is between B♭ and A and the eleventh is between F and F♯.

The figure is useful for finding the vibration ratios of the ordinary musical intervals. The major third, for instance, appears twice in the figure, as the interval between the 4th and 5th members of the series, and between the 8th and 10th. In each case the vibration ratio is 4 to 5.

There is a difference of practice amongst writers as to which member of the series is called the first harmonic. Some call the lowest note C the first harmonic, others give the name to the second term. We shall adopt the former plan, as it has the advantage of connecting the number of the harmonic with its vibration number. Thus if the fundamental has a frequency represented by 1, the fifth harmonic has frequency 5, and similarly for the others.

Another term sometimes applied to the terms of the series is *partials* or *harmonic partials*. When we have occasion to use this term we shall adopt the same convention as in the case of harmonics, and call the lowest note of the series the first partial. The term *partials* is not applied to *overtones*. It is restricted to the members of the harmonic series, and more particularly to such members as are present in a complex note, as will be explained in Chapter IX.

73. Effect of Imperfect Flexibility of a String.
The relations between the frequencies of a string, which we have deduced in this chapter, hold strictly only for a perfectly flexible string; that is, a string that can be bent without any force being needed.

Any actual string offers some slight resistance to bending, and the overtones consequently do not fall quite exactly into the harmonic series—each overtone is a little too high in pitch. For the lower members of the series, as given by a string that is not very short, the divergence from the harmonic series is not great.

CHAPTER IV

LONGITUDINAL WAVES

74. Transverse Waves cannot exist in a Gas.
The transverse waves discussed in Chapter III have not the same importance in the theory of Sound as have longitudinal waves, but we have dealt with them first, because they are more easily shewn experimentally. Everyone is familiar with the transverse waves on the surface of water, whilst longitudinal waves are visible only in exceptional cases, and their presence and properties must generally be demonstrated by indirect methods.

The importance of longitudinal waves arises mainly from the fact that air is the medium by which sound waves are generally carried from the place of their origin to the hearer, and no other waves than longitudinal waves can travel in air. A gas offers no permanent resistance to change of shape. Whilst the change of shape is taking place there will be some slight resistance in consequence of the viscosity and inertia of the gas. The resistance may be considerable when the change of shape of the gas takes place very rapidly, as when an aeroplane or motor car travels through the air at high speed. Such forces however are always opposed to the direction of motion, and disappear as soon as the motion ceases, so that there is no tendency for the gas to resume its original shape. The only change that gives rise to elastic forces is change of volume, and consequently the only waves that can be propagated through a gas are such as involve compressions and rarefactions. It is not possible to give a strictly logical investigation of the properties of longitudinal waves without introducing mathematics beyond the scope of this book, and many statements must be made the evidence for which cannot be given here.

We cannot *see* the waves in air, as we see the waves on a stretched string, and so we must for the present assume the possibility of their existence, leaving the experimental demonstration of their presence until we have discussed the phenomena of interference.

75. Experimental demonstration of Longitudinal Waves in a Spiral Spring. Waves closely resembling the longitudinal waves in air can be produced in a visible form in a horizontal spiral spring. The following dimensions and manner of supporting the spring are taken from Barton's *Text Book of Sound*, p. 7. " A helical coil should be wound in a lathe, the wire being of soft copper about 1·5 mm. diameter, its turns being 10 cm. diameter; their pitch, or distance apart longitudinally, may be about 1 cm., the whole coil being 2 m. long, thus containing about 200 turns. Each turn of the coil should be supported by a fine silk thread in the form of a V, each limb of the V being about 1 m. long, its two upper ends being fixed half a metre apart to a wood framework. These dimensions are chosen to insure a slow advance of the disturbance from one end to the other, and cannot without disadvantage be departed from at random."

If one end of the coil is pushed inwards a little, the turns near the end will be forced nearer together, forming what we may term a compression, and this compression will travel slowly to the other end of the coil. If on the other hand the end of the coil is pulled outwards a little, the turns near the end will be separated more widely from each other, and this state, which we may call a rarefaction, will also be seen to travel slowly along the coil. If now the end be first pushed inwards by say 1 in. from its equilibrium position, then drawn backwards 2 ins. and finally returned to its equilibrium position with a steady movement approximating to one complete harmonic vibration, one complete wave consisting of a region of condensation followed by a region of rarefaction will be seen to travel along the coil. If any one ring of the coil is watched whilst the wave is passing over it, that ring will be seen to perform a single vibration similar to the vibration that the end was made to perform, and will return to rest as soon as the wave has passed. Thus, if the left-hand end of the coil is made to perform a harmonic vibration, each ring in turn will perform a similar vibration, and each ring will begin its vibration a little later than its left-hand neighbour.

The motion of the successive turns of the coil is very similar to what can be shewn theoretically to be the motion of successive layers of air over which a sound wave is passing.

76. General description of the propagation of Air-Waves. Having thus obtained a general idea of the nature of longitudinal wave motion, we pass to the case of waves in air.

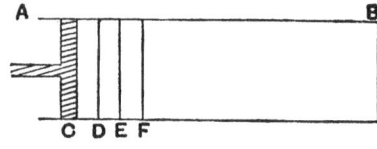

Fig. 32

Imagine a circular cylinder AB containing near one end a piston C, which fits the cylinder air-tight, but can move easily within it. Suppose the piston is moved rapidly a little to the right, and is held at rest in its new position. A thin layer of air CD close to the piston will be momentarily compressed, whilst the rest of the air in the cylinder will be momentarily unaffected, for the air has inertia, and the motion of the layer near C cannot set the next layer in motion instantaneously. The layer CD will therefore be compressed, and its pressure will be greater than that of DE; but this state of affairs cannot remain, and CD will at once begin to expand towards the right, communicating its compression to DE, which in turn will communicate it to EF, and so on, a pulse of compression travelling onwards towards the right. We will for the moment postpone consideration of what happens when the pulse reaches B.

The foregoing description of the propagation of a pulse of compression is incomplete, as it does not shew whether the layer CD is left partly compressed, rarefied, or in its original state, when the pulse has passed away to the right. The motion of the piston not only compresses the layer CD, but also sets it in motion, and it can be shewn mathematically that when the layer CD has reduced itself to rest by communicating its momentum to DE, it has also got rid of its compression, and has therefore returned completely to its original state. The same is true of each succeeding layer, and therefore when the pulse has passed, the air is left in its undisturbed state.

77. Experimental Analogies. The case is analogous to that of a single transverse wave sent along a stretched string. We saw in the experiment with an indiarubber cord described in Chapter III, that as soon as the wave had passed any section of the cord, that section was left undisturbed. The whole of the energy of the wave was transferred from section to section leaving none behind, except such small amount as was lost by conversion into heat in consequence of the imperfect elasticity of the string.

The following experiment is still more closely analogous to the propagation of an air-wave. Place a dozen glass or steel balls in contact with each other on a board in which a V groove has been cut to keep the balls in line. Now roll another ball along the groove so as to strike one end of the line. It will be found that the ball at the other end will leave the line and roll away, leaving the remaining balls at rest. The energy of the first ball has been transferred completely from ball to ball, until it reaches the last, and this last ball, finding no other to which it can transfer its energy, is set in motion.

78. Longitudinal Vibrations of Particles of Air. Let us return now to the cylinder and piston. Instead of imagining the piston to be moved quickly to the right, let us suppose it is moved to the left. The layer of air CD will then be rarefied, and its pressure will fall. Air will flow in from DE to restore the pressure of CD, and DE will be rarefied, and the same will happen to each succeeding layer. Thus a pulse of rarefaction will travel to the right.

If now the piston is kept vibrating continuously, a train of waves consisting of alternate condensations and rarefactions will travel along the tube.

It is evident that the particles of air in contact with the piston must follow its motion, and vibrate in a direction parallel to the axis of the tube. This causes a similar motion in the next layer and in all the succeeding layers, so that every particle of air vibrates to and fro in the direction in which the wave travels. It is for this reason that waves of condensation and rarefaction are called *Longitudinal Waves*, to distinguish them from waves such as those described in Chapter III, where the individual particles vibrate in lines transverse to the direction in which the waves travel. In the case of transverse waves one complete wave includes a crest and a trough. In the case of longitudinal waves one wave consists of a region of condensation and a region of rarefaction.

79. Properties of Longitudinal Waves. It can be shewn that, if the piston is made to vibrate harmonically, every particle of air along the tube will vibrate harmonically as the waves pass over it. Further, if we consider a row of particles lying along a line parallel to the axis of the tube, and equally spaced along that line when no waves are passing, there will be a uniform retardation of phase from particle to particle, as we pass along the series from left to right, when the waves are made to travel from left to right. A train of waves would result from *any* periodic vibration of the piston, whether harmonic or not, but the waves resulting from harmonic vibrations are the simplest theoretically and, as we shall see when we discuss Fourier's Theorem, they may be regarded as the fundamental form of wave motion, from which all other forms of waves can be derived. We shall limit ourselves to harmonic waves for the present.

If we assume that each particle of the series mentioned above vibrates harmonically, with a uniform retardation of phase as we pass from particle to particle along the series, we can deduce many of the more important properties of longitudinal waves.

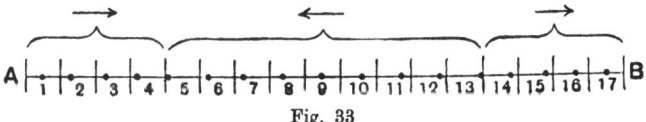

Fig. 33

Let AB be the line along which the particles lie, and

62   LONGITUDINAL WAVES   [CH. IV

suppose that when no waves are passing there is one particle in the centre of each of the 17 sections into which the line is divided by the short vertical lines. When the waves are passing, the particles will perform harmonic vibrations. Let us suppose that the extreme range of vibration of a particle is the length of one of the sections, and each particle is a sixteenth of a period behind its left-hand neighbour. Suppose particle 1 is passing from left to right through its equilibrium position at the moment we are considering; then particle 2 will also be moving from left to right, but will not reach its equilibrium position until one sixteenth of a period later. Similarly particle 3 will be one eighth of a period behind, 4 will be three sixteenths behind, and 5, being a quarter period behind, will be just at the left-hand end of its swing, and so momentarily at rest. Particle 6 will be coming up to its left-hand limit, and so will be moving to the left.

It is easy to follow out this process along the row, and it will be seen that all the particles from 1 to 4 and from 14 to 17 are moving towards the right, and all from 6 to 12 towards the left. 5 and 13 are at the ends of their swing, and 1, 9 and 17 are passing through their equilibrium positions.

As the particles are all vibrating harmonically, we know that they have their greatest velocity when passing through their equilibrium positions, and have no velocity at the ends of their swings. The velocity is therefore a maximum at 1 and diminishes to zero at 5. It is then reversed and increases to a maximum at 9, and again diminishes to zero at 13, where it is once more reversed, and increases to a maximum at 17.

80. Condensation and Rarefaction. We must next find the positions of the regions of condensation and rarefaction. Each particle is in advance of its right-hand neighbour by one sixteenth of a period; that is, the *time* difference is constant for each pair of neighbours. The *space* difference however will vary, for a particle moves faster the nearer it is to its equilibrium position. Suppose for the sake of argument that the period of vibration is one second, and the range of vibration from end to end is one inch. (Ordinary sound waves in air have a much smaller period and amplitude, but this does not affect the principle of the method.)

We can find graphically by use of a diagram similar to Fig. 7 the distance a particle moves in each sixteenth of a second. Suppose a radius moves uniformly round O in a

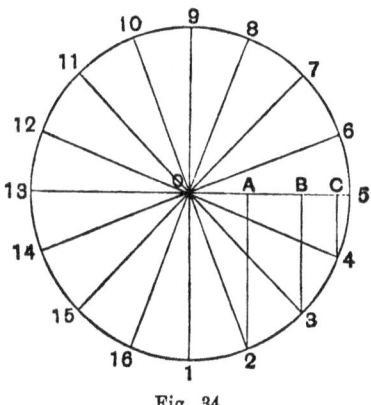

Fig. 34

counter-clockwise direction, making one complete turn in a second. Let us reckon the time from the moment when the radius is in the position 1, and drop perpendiculars on the horizontal diameter, so that the position 1 corresponds to a particle passing from left to right through its equilibrium position. One sixteenth of a second later the radius is at 2, and the particle is at a distance OA from its equilibrium position. After another sixteenth of a second it is at B, and so on. It will be found by measurement* that the lengths OA, AB, BC and CD, which are the distances moved through in consecutive sixteenths of a second, are ·19, ·16, ·11, ·04 inches respectively, and the displacements OA, OB, OC and OD are ·19, ·35, ·46 and ·50 respectively.

Returning now to Fig. 33: 1 is in its equilibrium position, 2 is ·19 in. away from its equilibrium position and hence 1 and 2 are ·81 in. apart, 3 is ·35 in. from its equilibrium position, and

* The lengths OA, OB and OC can be found at once from a table of cosines, for it is evident that $OA = \frac{1}{2} \cos 67\frac{1}{2}°$, $OB = \frac{1}{2} \cos 45°$ and so on.

hence 2 and 3 are ·84 apart. Similarly 3 and 4 are ·89 apart, and 4 and 5 are ·96 apart.

When no waves are passing and the air is at rest, any two neighbouring particles are 1 in. apart. At the moment under consideration all the pairs of particles from 1 to 5 are less than an inch apart, and consequently there is condensation in this region; the condensation being greatest at 1 and diminishing as we go towards 5. Similarly it can be shewn that all the pairs of particles from 5 to 13 are more than an inch apart, the excess being greatest at 9. Consequently there is rarefaction in this region with a maximum at 9. At 13 the rarefaction changes again to condensation and so on.

81. Summary of the Properties of Longitudinal Waves. We can summarize our conclusions thus:

(1) When a particle is passing through its equilibrium position it has its maximum velocity.

(2) If it is then moving in the same direction as the wave is travelling, it is at the centre of a region of condensation; if it is moving in the opposite direction to that in which the wave is travelling, it is at the centre of a region of rarefaction.

(3) When a particle is at either end of its swing it has no velocity, and the air in its immediate neighbourhood has its normal density, with condensation on one side and rarefaction on the other.

82. Associated Curves. By making use of certain conventions we can shew these relations by curves. Let an ordinate drawn upwards represent the displacement in the direction of the motion of the wave of a particle whose equilibrium position is at the foot of the ordinate, and vice versa. We shall then get a curve such as that shewn in Fig. 35 for the series of particles in Fig. 33.

Fig. 35. Displacement Curve.

To shew the changes of velocity, let velocities in the direction in which the wave travels be represented by upward ordinates and velocities in the opposite direction by downward ordinates. We then get the curve of Fig. 36.

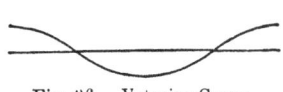

Fig. 36. Velocity Curve.

Lastly, let upward ordinates represent condensations and downward ordinates rarefactions, and we get the curve of Fig. 37.

These curves are all sine curves, but the velocity and condensation curves are moved a quarter of a wave-length to the right as compared with the displacement curve.

Fig. 37. Condensation Curve

It is important to remember that the displacement curve does not shew directly the displacements in the directions in which they actually take place, but merely represents them conventionally. It is for this reason often called the *Associated Curve* of the wave.

83. Sound Vibrations are superposed on Molecular Motions. A further warning should be given here. The "particles" of which we have spoken must not be taken to be individual molecules of the gas, but small volumes each containing a large number of molecules. According to the Kinetic Theory of Gases each molecule is in irregular and rapid motion, even though the gas as a whole is at rest. The elastic vibrations of which we have spoken are regular periodic motions, which are superposed on the irregular motions of the molecules. If we take as our "particles" small cubes with edges of, say, ·001 millimetre, each cube will contain so many molecules that its centre of gravity may be taken to be at rest, when the gas as a whole is undisturbed; and yet the cube will be so small that we need not consider the difference between the displacements of two opposite faces, but can regard it as moving bodily to and fro without change of shape or size.

84. Crova's Disc. The conclusions at which we have arrived can be seen to hold for the waves in the spiral wire described on p. 58. A still simpler piece of apparatus for shewing vibrations similar to those in a longitudinal wave is Crova's Disc. To construct such a disc proceed in the following way. In the centre of a large card draw a small circle, say $\frac{3}{8}$ in. in diameter. Divide the circumference of this circle into 8 equal parts, and number the points of division 1, 2, 3, etc. With 1 as centre draw a circle of $\frac{1}{2}$ in. radius, with 2 as centre draw a circle of $\frac{3}{4}$ in. radius, with 3 as centre draw a circle of 1 in. radius and so on, at each step increasing the radius by $\frac{1}{4}$ in., and taking the points on the circumference of the small circle in order as centres. Draw as many circles as the card will hold, going as many times round the circle of centres as may be necessary.

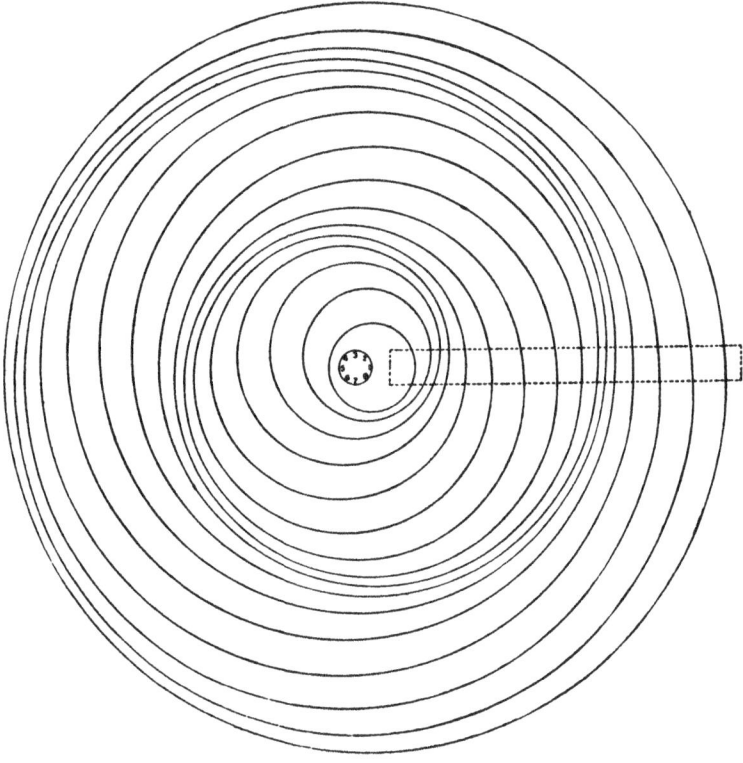

Fig. 38

We shall then have a diagram similar to Fig. 38. Put a pin through the centre of the card and hold before it another card with a slit cut in it, the slit being in the position shewn by the dotted lines in the figure.

If now the card with the circles drawn on it is made to rotate round the pin as centre, the small pieces of the circles seen through the slit will be seen to vibrate to and fro, and waves of compression and rarefaction will travel along the slit.

85. Wave-Fronts. We have hitherto discussed only the nature of sound waves in a cylindrical tube. The thin sheet of compression produced by a rapid movement of the piston of Fig. 32 is a plane section at right angles to the axis of the tube, and it remains so as it travels along. If a surface be drawn in such a way that at any moment all the particles of air in the surface are in the same phase, the surface is called a *Wave-Front*. There are an infinite number of wave-fronts even in a single wave-length. The wave-front might for instance be the locus of all the adjoining points that are at the moment at the point of maximum compression, or that are passing in a particular direction through their equilibrium position, or are in any other phase we may choose to select.

It is evident that the wave-fronts for the air-waves we have considered so far are planes, and the waves are therefore called *Plane Waves*. Such waves are exceptional, and we must now pass to the more usual type of waves, where the wave-fronts are in general not plane.

86. Spherical Waves in Air. Imagine that we have at some point in the air a small sphere which expands and contracts rapidly. When it expands, a spherical layer of air near its surface is compressed, and the compression spreads outwards. When it contracts the layer of air near its surface is rarefied, and the rarefaction spreads outwards in the rear of the compression. If the sphere continues to expand and contract periodically, we shall have a series of sheets of compression and rarefaction spreading out alternately. As there is no reason why a condensation or rarefaction should travel faster in one direction than in another, it may be assumed that it will travel with the same velocity along any straight line radiating out from the small sphere. The wave-front in contact with the small sphere is spherical, and therefore every other wave-front is also a sphere with the centre of the pulsating sphere as centre. Further, it is clear from symmetry that the direction of vibration of any particle of air is along a radius of the system of spherical wave-fronts.

87. Direction of Propagation of a Wave. We may borrow a convenient expression from Optics, and say that the pulsating sphere sends out *Rays of Sound* in every direction, and the direction of propagation of the sound at any point is along the ray through that point; or, since the rays are radii of the wave-fronts, the direction of propagation at any point is at right angles to the wave-front passing through that point. This is an instance of the general law that the direction of propagation of any wave motion at any point is normal to the wave-front passing through that point, whatever the shape of the wave-front may be. We shall have occasion to make use of this law in a later chapter.

88. Energy of Air Waves. As the sphere vibrates, it meets with resistance from the air, and therefore work has to be done to make it vibrate. Energy is thus communicated to the air, and travels outwards with the waves. The energy of the waves is partly kinetic and partly potential. A particle executes harmonic vibrations about its mean position, as the waves pass over it, and it therefore possesses kinetic energy, except at the moment when it is at the end of its swing. We have seen also that there are regions of condensation and rarefaction in a wave, and therefore there is potential energy everywhere, except at a point where we are passing from a condensation to a rarefaction, or vice versa. The energy is not distributed uniformly over the wave, for we saw that at the points of no velocity there is also no compression, and consequently at these points the air has no energy above what it would have if no waves were passing.

It can be shewn that the total energy per unit of volume of a medium through which sound waves are passing is proportional to the square of the amplitude of the waves, and inversely proportional to the square of the wave-length; and further, that the energy in a whole wave is at every instant half kinetic and half potential.

89. Constancy of the Intensity of Plane Waves.

Plane progressive waves such as those travelling along a tube cannot spread out sideways, but are confined to the cylindrical column of air. The amplitude remains constant, and the flow of energy is the same through any cross section of the tube. The amplitude is always constant when the wave-fronts are plane, whether the sound is confined in a tube or not. We shall see in a later chapter that it is possible by reflection at a curved mirror to convert a diverging beam of sound in the open air into a parallel beam with plane wave-fronts, and so enable it to travel over a long distance without loss of energy. The case is analogous to that of a search light, where the diverging rays of light from an electric arc are converted into a parallel beam of plane waves by a mirror, for the same reason.

90. Change of Intensity of Spherical Waves.

When sound waves spread out in every direction from their source, it is clear that their intensity must fall off as they spread outwards. The total amount of energy crossing any spherical shell with the source as centre must be the same, whilst the area of the shell is proportional to the square of its radius. Hence the quantity of energy passing per second through a square centimetre of the surface of any shell is inversely proportional to the square of the radius of the shell, or the intensity of sound varies inversely as the square of the distance from the source, when the wave-fronts are spherical. It is evident that the law still holds, even though the wave-fronts are not complete spheres. It holds, for instance, for sound travelling from the apex towards the wide end of a hollow cone. If the sound travels from the wide to the narrow end of the cone the amplitude of the waves increases, and the intensity of the sound increases. This is the principle of the ear-trumpet used by deaf persons.

91. Reflection at the closed end of a tube.

Let us next consider what happens when a pulse of compression reaches the closed right hand end of the tube of Fig. 32. We have seen that the pulse travels along, leaving the air behind it at its normal pressure. When the pulse reaches the closed end, the motion of the particles is checked, and we have for a moment a thin layer of compressed air at rest in contact with the end, and the rest of the air in its normal state. This is exactly what we had at the left hand end when the piston was suddenly pushed in a little and then held at rest. Consequently what was said as to the result of pushing in the piston applies in this case, and the pulse of compression runs back to the left. Thus we see that a pulse of compression is reflected from a closed end, and returns as a compression. Similarly a pulse of rarefaction is reflected as a rarefaction, and a continuous train of waves arriving at the closed end will be reflected and return as a similar train, with no change of phase at the end where the reflection takes place. This agreement of phase in the direct and reflected waves should be specially noted, for we shall see presently that if the right hand end of the tube is open, there will still be reflection, but there will be a change of phase of half a period.

In the case of longitudinal waves it can be shewn that, if the change of pressure is proportional to the change of volume of the air, we can find the actual displacement at any point, when both the direct and reflected waves are passing, by adding the displacements due to the two, taking account of their signs as we did with transverse waves. As we have already mentioned, the change of pressure is not strictly proportional to the change of volume for air, but it is very nearly so if the displacements are small, as they are for ordinary sound waves, and as a first approximation we can assume the proportionality to hold. In a later chapter we shall discuss some results of the want of strict proportionality. The same method can be used for compounding two compression curves or two velocity curves, but it must be remembered that we are treating here only of trains of waves travelling in the same straight line either in the same or in opposite directions. We shall discuss the case of two trains of waves crossing each other at any angle in Chapter VII.

92. Stationary Vibrations in Air. It is not necessary to enter again into a complete description of the way in which stationary waves arise from the superposition of the direct and reflected trains, as most of what was said in Chapter III about the reflection of transverse waves applies here also. Fig. 27 will apply to air waves if we regard the curves as the associated displacement curves for the air waves.

91–93] LONGITUDINAL WAVES 71

It is clear that there cannot be any displacement at the closed end of the pipe, and this will be secured if the direct and reflected trains always give equal and opposite displacements at B, so that the resultant displacement at that point is always zero.

The curves of Fig. 27 give this relation between the displacements due to the two trains at B, and it follows as in Chapter III that there will be a series of points half a wavelength apart at which the displacement is always zero, one of these points being at the closed end of the pipe. We have as before a train of stationary waves consisting of vibrating segments separated from each other by equidistant nodes.

*93. Distinction between Transverse Wave curves and the associated curves of Longitudinal Waves. There is a point of difference between the cases where the curves of Fig. 27 are applied to transverse and longitudinal waves, which might present a difficulty to the student.

The curves have not the same meaning in the two cases. The curve for a string shews the actual shape of the string, the direction and length of any ordinate shewing directly the direction and amount of displacement of the small section of the string, which would be situated at the foot of the ordinate, if no wave were passing. The meaning of the curve is therefore the same for the direct and the reflected wave. In the case of the curve for air waves, the meaning of an ordinate was defined with reference to the direction in which the wave travels; an upward ordinate denoting a displacement in the direction of motion of the wave and vice versa. The direct wave is travelling to the right and the reflected wave to the left, and therefore if we were to adhere to our convention, an upward ordinate of the direct wave would mean a displacement to the right, and an upward ordinate of the reflected wave would mean a displacement to the left. Since the curves are used to find the resultant displacement by compounding the components, it would be inconvenient to have these different meanings for the ordinates of the two curves, and we therefore make the convention that in each case upward ordinates are to represent displacements to the right, regardless of the direction of motion of the wave. We saw that when the wave is travelling to the right, those air particles which are moving to the right are situated in a region of compression. If the curve of Fig. 39 is the associated displacement curve for such a wave, it will be seen that the ends of the ordinates near B are moving upwards, and so shew the displacements of particles moving to the right.

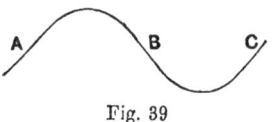

Fig. 39

Hence the front face of each crest corresponds to a region of compression; and it will be seen similarly that the rear face A represents a region of rarefaction. This relation is true for the direct wave, but not for the reflected wave, where an upward ordinate represents a displacement contrary to the direction of the wave. In this case the wave is travelling to the left, and therefore the compression is situated where the particles are moving to the left, that is, where the ends of the ordinates are moving downwards. Thus it follows that if the curve of Fig. 39 is moving to the left, B is still a region of compression and A a region of rarefaction, but the rarefaction is now on the front face of the crest, and the compression on the rear face. Hence we see that in Fig. 27 the displacements at B due to the direct and reflected waves are equal and opposite, and at the moment for which the figure is drawn both D and E are situated in regions of rarefaction. Consequently we get a rarefaction and no displacement at B as the result of compounding the direct and reflected waves at that point.

94. Properties of stationary Longitudinal Vibrations. The Associated Curve for stationary air vibrations is a sine curve, and passes through the same series of changes as the curve for stationary transverse vibrations.

Fig. 40 is the associated curve for three vibrating segments separated by nodes at A and B.

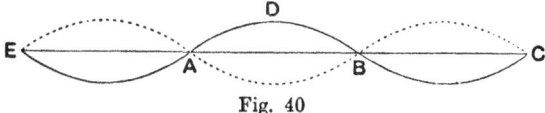

Fig. 40

The curve is drawn in two extreme positions—those in which the particle midway between the nodes A and B is at the extreme right hand of its swing and the extreme left hand respectively. The former is shewn by a full line and the latter by a dotted line. The state of affairs represented by the dotted curve is reached half a period after that represented by the full curve.

At the moment shewn by the full curve every particle is at the end of its swing, and hence is momentarily at rest. All the particles between A and B, having ordinates above the axis, are at the moment displaced towards the right. All those between B and C are displaced towards the left. It follows that there is a crowding in of the particles towards B from both sides, or B is the centre of a region of condensation. Similarly there is a drawing away of the particles from A, which is consequently at the centre of a region of rarefaction. At the point D, half way between the nodes, the tangent to the curve is horizontal. There is consequently a small range at this point where all the particles have the same displacement. That is to say, they are the same distance apart as when there are no waves, and there is neither condensation nor rarefaction at D. As the full curve moves downwards, the tangent at D is always horizontal, and therefore there is never any condensation or rarefaction at points half way between two consecutive nodes. There will be a moment when the curve takes the form of the straight line EC. At this moment each particle is passing through its equilibrium position. There is nowhere compression or rarefaction, but all the particles have their maximum velocity—to the left between A and B and to the right between B and C. A moment later the curve has got below the axis between A and B and above between B and C. The air is now displaced away from B on both sides and therefore B is now a centre of rarefaction, and A and C are centres of compression.

The nomenclature for stationary air vibrations is the same as that given on p. 49 for stationary vibrations on strings.

95. Summary of the properties of Stationary Vibrations in Air. We can now summarize the chief properties of stationary waves in the following statements:

(1) There is never any displacement at a node.

(2) At any moment the displacement is greater at an antinode than at any other point.

(3) At any moment the velocity of the air is greater at an antinode than at any other point.

(4) As we pass along a series of vibrating segments, the displacement is alternately to the right and to the left in the consecutive segments.

(5) At an antinode the air always has its normal density.

(6) A node is a centre of maximum compression and maximum rarefaction alternately.

(7) As we pass along a series of nodes we find them centres of compression and rarefaction alternately.

If the spiral spring described at the beginning of this chapter has its two ends fixed, it can be made to perform stationary vibrations. Take the spring in the hands at two points each of which is about a quarter of the length of the spring from an end. Draw these two points of the spring apart by separating the hands a little farther from each other, and then release the spring. It will perform stationary vibrations with a node at each end and one in the middle, and the statements made above can be verified by observation of the motion.

96. Comparison of Stationary Vibrations and Progressive Waves. It should be noticed that the relations of velocity and condensation to displacement are not the same for stationary vibrations as they are for progressive waves. Figs. 41 and 42 shew the associated curves for the two cases.

In Fig. 42 the air particles are assumed to be neither at the end of the swing nor passing through the equilibrium position, but in some intermediate position, and moving *towards* their equilibrium position. If they were at the ends of their swings, they would be momentarily at rest and the velocity curve would be a straight line, whilst the condensation curve would have its greatest amplitude. If they were passing through their equilibrium positions, the condensation curve would be a straight line, and the velocity curve would have its maximum amplitude. There is of course the further difference between the two sets of curves that those in Fig. 41 travel onwards with the wave, whilst those in Fig. 42 retain the same positions, and merely change in amplitude.

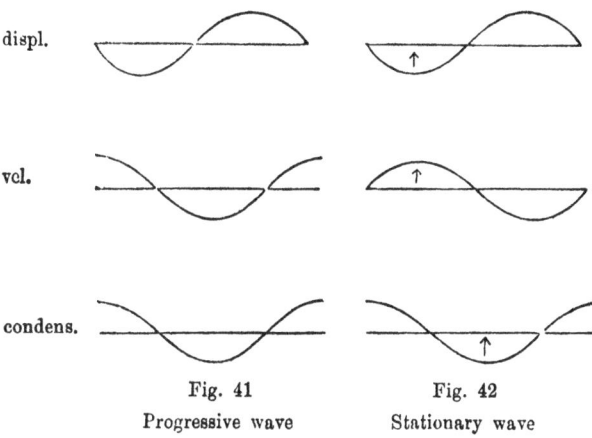

Fig. 41 Fig. 42
Progressive wave Stationary wave

The distribution of energy is not the same in a stationary wave as in a progressive wave. In the case of a series of stationary waves there is no flow of energy along the series. The ratio of the kinetic to the potential energy is the same at every point at any moment, but the ratio varies from moment to moment. When, for instance, the particles are at the extreme end of a swing—which occurs at the same moment everywhere—the energy is everywhere wholly potential, and when the particles are moving through their equilibrium positions, the energy is everywhere wholly kinetic.

97. Reflection at the open end of a Tube. We must next return to the cylinder and piston of Fig. 32, and find in what way reflection takes place when the right hand end is open.

When a pulse of compression is passing over any layer of air, the layer has both kinetic and potential energy. It is compressed, and it has a velocity towards the right. It transfers the whole of its energy to its right hand neighbour, and is itself brought to rest with no compression and no velocity remaining. This transference of energy continues unchanged along the tube, so long as the resistance to compression of each succeeding layer is the same. At the open end however the circumstances are different. Inside the tube a layer acted on by a compressed layer behind it can only move forwards, whilst the layer at the open end can also spread sideways. Hence a compressed layer at the open end meets with less resistance to expansion than it would meet with inside the tube, and in expanding to such an extent as will reduce its compression to zero, it does not use up the whole of its energy, but has still some kinetic energy left. Consequently the air, instead of coming to rest when it reaches its normal density, overshoots the end, and comes to rest only when a rarefaction has been produced. This rarefaction will cause a pulse of rarefaction to run down the tube, whence we see that when a pulse of compression is reflected from an open end, it changes its sign and returns as a rarefaction.

There is another difference between the reflections at a closed and open end which should be mentioned here, though it has no effect on the object we have in view at present. When the reflection is from an open end it is only partial, as part of the energy of the direct wave travels out into the open air.

98. Effect of a change in the bore of a tube.

It is clear from the explanation we have given of the reflection at an open end, that there would be some reflection if a pulse travelled along a tube which had a sudden change of diameter at some point.

If a compression starts from A (Fig. 43) and travels to the right, it meets with lessened resistance on reaching B, overruns itself, and part of its energy is reflected back to A in a rarefaction. If the compression starts from C and travels towards the left, it meets a greater resistance on reaching B, and part of its energy is reflected back to C in a compression.

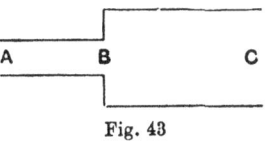

Fig. 43

In each case part of the energy goes on and part is reflected, and the amount reflected depends on the relative cross sections of the two parts of the tube. If there is little change of section there is little reflection.

99. Effect of a change in the density of the medium. We should also have some reflection at the surface separating the two gases, if part of the tube were filled with one gas and part with another. Suppose for instance that the tube is held vertically and the lower half is filled with carbon dioxide and the upper half with air. A pulse of compression travelling down the tube will be partially reflected as a compression at the dividing surface, for, though the elastic resistance of the carbon dioxide is not much different from that of air, its inertia is about half as great again, and consequently the pulse will meet with greater resistance, when it reaches the surface of separation. If the pulse travels upwards through the carbon dioxide, there will be partial reflection again, but with change of sign.

100. Stationary Vibrations in a pipe with an open end. As a compression is reflected as a rarefaction and a rarefaction as a compression at the open end of a pipe, it follows that a train of waves of alternate compressions and rarefactions will be reflected as a similar train; but, as compared with a closed pipe, the reflected train will be shifted backwards through half a wave-length in consequence of the change of phase at the moment of reflection. The result of compounding the direct and reflected trains will as before be a series of stationary waves. If Fig. 27 be redrawn with the reflected train moved backwards or forwards half a wavelength—it does not matter which—it will be found that the open end is not now a node but an antinode, and the first node is a quarter wave-length from the end.

This statement is not strictly accurate. It is an approximation which is farther from the truth the greater the diameter of the pipe. The problem of finding the exact circumstances of the reflection at the open end of a pipe is difficult, and has not yet been fully solved. It appears that the effect is the same as if the reflection took place a little distance beyond the end, or as if the reflection took place at the end, but the pipe were longer by an amount called the *correction for the open end*. Experiments have shewn that for a circular pipe with thin walls the correction is about three-fifths of the radius of the pipe. We shall return to this subject when we discuss the properties of organ pipes.

CHAPTER V

VELOCITY OF LONGITUDINAL WAVES

101. Velocity of Waves in Air. If we assume that longitudinal waves can be propagated unchanged in air, we can shew that the velocity of the waves must be $\sqrt{\dfrac{E}{\rho}}$. In this expression E is the coefficient of volume elasticity, or the ratio of a small increase of pressure to the small diminution of volume per unit volume that results from that increase of pressure, and ρ is the density of the gas, or the mass of unit volume when the gas is in its normal state.

*Suppose sound is travelling against a wind which blows with the same velocity as that with which sound travels in air, then, though the waves are moving relatively to the air, they are at rest relatively to the ground. Hence the velocity, pressure and density of the air at any point fixed relatively to the ground remain constant. Imagine two unit areas A and B fixed relatively to the ground, and so situated that the line joining their centres is in the direction in which the wind blows, their planes being perpendicular to this line. Suppose A is at a part of a particular wave where the air has its normal density, and B is at some other part of the wave, and suppose the wind blows from A to B. Let us suppose for the sake of definiteness that B is situated in a region of rarefaction, where the pressure is less than at A. It is clear that the same mass of air must pass per second through A and B, as the state of the air as regards density and velocity remains constant at any point fixed relatively to the ground and there cannot be any change in the mass of air between the two areas. Further, since the volume of a given mass of air depends on its pressure, a greater volume of air must flow through B than through A, or the velocity of the air is greater at B than at A.

Let V, p, ρ be the velocity, pressure and density respectively at A, and V', p', ρ' be the same quantities at B.

We have then the two following conditions:

(a) The mass of air between the planes is invariable, and therefore the mass of air crossing the plane A per second is equal to the mass crossing B per second. This gives the equation
$$V\rho = V'\rho' \quad \dots\dots\dots\dots\dots\dots\dots(1).$$

(b) The velocity at B being greater than that at A the air must have gained momentum between A and B, and by the Second Law of Motion the external force on the layer of air which is between A and B at any instant is equal to the rate of change of momentum of this same mass of air at the same instant, whence
$$p - p' = V'\rho' \times V' - V\rho \times V \quad \dots\dots\dots\dots(2).$$

Substituting in (2) the value of V' given by (1) we have

$$p - p' = V^2 \rho \frac{\rho - \rho'}{\rho'}$$

or
$$V^2 = \frac{\rho'}{\rho} \frac{p - p'}{\rho - \rho'}.$$

From the definition of the coefficient of volume-elasticity we may write
$$E = \frac{p - p'}{(v' - v)/v},$$

where v is the volume of unit mass, or the reciprocal of ρ. Substituting $\frac{1}{\rho}$ for v we have

$$E = \frac{p - p'}{\rho - \rho'} \rho',$$

and therefore
$$V^2 = \frac{E}{\rho}.$$

Now V is the velocity at A, where the air has its normal density, and is therefore the general velocity of the wind, which was chosen so as to be equal and opposite to the velocity with which the wave would travel in still air. Consequently the velocity of sound waves through the air is $\sqrt{\dfrac{E}{\rho}}$.

102. Velocity of Elastic Waves of any type.

It will be seen that this method is analogous to that given in Chapter II for finding the velocity of a transverse wave along a string, and the resulting expression is similar. When a wave of any kind is propagated by the action of elastic forces, the velocity of the wave is proportional to the square root of the appropriate coefficient of elasticity, and inversely proportional to the square root of the appropriate expression for the inertia of the system.

If, for instance, one end of a long cylindrical rod be rapidly twisted a little, first to the right and then to the left, a torsional wave will run along the rod, each section of the rod in turn twisting first to the right and then to the left. The velocity of such a wave is proportional to the square root of the modulus of torsion of the rod, and inversely proportional to the moment of inertia of unit length of the rod about its axis of figure.

103. Newton's Calculation of the Velocity of Sound.

The expression $\sqrt{\dfrac{E}{\rho}}$ for the velocity of sound in a gas was first obtained by Newton, who used it to calculate the velocity in air from the measured values of the coefficient of elasticity and the density. He assumed that Boyle's Law holds for the compressions and rarefactions in a sound wave, namely that pv is constant. We can shew that $E = p$ if Boyle's law holds. For suppose the pressure is increased from p to $p + \delta p$, where δp is small, and v is in consequence changed to $v - \delta v$; then

$$pv = (p + \delta p)(v - \delta v)$$
$$= pv + v\,\delta p - p\,\delta v - \delta p\,\delta v.$$

The last term being the product of two small quantities may be neglected, and we have therefore $v\,\delta p = p\,\delta v$, or

$$p = v\frac{\delta p}{\delta v},$$

which is by our definition the value of E. This is called the Isothermal Coefficient of Elasticity of a gas, for Boyle's Law assumes that the temperature is constant.

Thus the velocity of sound is $\sqrt{\dfrac{p}{\rho}}$ if Boyle's Law holds for the compressions and rarefactions. If v be the volume of unit mass, we have $v = 1/\rho$ and therefore the expression for the velocity can be written $\sqrt{pv}$.

Newton, using the values accepted at that time for p and ρ, found 979 feet per second for the velocity of sound in air, whereas the real velocity was known to him to be about 1100 feet per second, and the discrepancy would have been still greater if he had used more accurate values for p and ρ. He accounted for the discrepancy of 1 part in 8 by assuming that one-eighth of the space travelled over by the sound was occupied by the particles of air, and that the formula applied only to the remaining seven-eighths, the sound passing instantaneously through the actual particles. He also assumed that any water vapour that might be present had no share in conducting the sound. These were unwarranted assumptions with no experimental justification, and it was a century before the true explanation of the discrepancy was found by Laplace.

104. Laplace's correction of Newton's calculation. When a gas is compressed it is heated, as is shewn, for instance, by the considerable rise of temperature of a bicycle pump after it has been used for inflating a tyre. Similarly, when a gas is rarefied it is cooled.

Now Boyle's law is true only when the temperature of the gas is kept constant, and therefore, if it is to be applicable to sound waves, we must suppose that the heat produced in the regions of compression is transferred so rapidly by radiation and conduction to the regions of rarefaction, that the gas is maintained constantly at the same temperature everywhere.

In view of the rapid changes of pressure in air when sound waves are passing, and the poor radiating and conducting powers of gases, this equalization of temperature seems improbable, and Laplace pointed out that we should use not the Isothermal but the Adiabatic Coefficient, which is the value of $v\dfrac{\delta p}{\delta v}$ when no heat is allowed to enter or leave the air. It is evidently greater than the Isothermal Coefficient, for the pressure of a gas can be raised either by heating the gas without change of volume, or by compressing it without change of temperature. In the case of an adiabatic compression the volume is reduced, and also the temperature is raised, since the heat produced by the compression is not allowed to escape. Hence the rise of pressure is greater than it is when the same compression is made isothermally, for in this latter case the heat produced is removed, and plays no part in raising the pressure.

105. Adiabatic Coefficient of Elasticity of a gas. It can be shewn by thermodynamical methods that when the changes of pressure are adiabatic pv^γ is constant, where γ is constant for any one gas, but not the same for all gases.

If we give the same meanings to the letters as we did when finding the isothermal coefficient of elasticity, we have, when the compressions are adiabatic,

$$pv^\gamma = (p + \delta p)(v - \delta v)^\gamma.$$

The last factor on the right can be written $v^\gamma \left(1 - \dfrac{\delta v}{v}\right)^\gamma$
Expanding this by the Binominal Theorem and neglecting all terms containing higher powers of δv than the first we have

$$pv^\gamma = (p + \delta p)\, v^\gamma \left(1 - \gamma\, \dfrac{\delta v}{v}\right).$$

Multiply out, neglecting the product of the small quantities δp and δv, and we find

$$\dfrac{v\, \delta p}{\delta v} = \gamma p.$$

Thus we see that the adiabatic coefficient of elasticity is γ times the isothermal coefficient.

It can be shewn that the specific heat of a gas when the pressure is kept constant is γ times the specific heat when the volume is kept constant. It is from this relation that γ is usually called the *Ratio of the Specific Heats*.

The constant γ is of importance in the theory of gases, and many experiments have been made to find its value. It has been found to have the value 1·66 for monatomic gases such as argon and mercury vapour, 1·41 for most diatomic gases, including air, oxygen, nitrogen and hydrogen, and lower values for gases of greater atomicity. Most of the determinations have been made by measuring the velocity of sound in a gas, and thence calculating γ. It would not, of course, be legitimate to use a value of γ so determined for testing the correctness of the formula $V = \sqrt{\dfrac{\gamma p}{\rho}}$, but there are other methods such as that of Clément and Desormes, which are independent of the velocity of sound, and it is found from these that γ has the value 1·41 for air.

When Newton's formula is corrected by replacing p by γp, it gives a value for the velocity of sound which is in agreement with the results of experiment.

It might be thought that though there may not be enough transference of heat from the hot to the cold places to keep the temperature constant, there might be some transference, and that we ought therefore to use a value for the coefficient of elasticity intermediate between p and γp. It has been shewn however by Stokes that, if there were any appreciable transference of heat short of what is required to make the changes isothermal, the sound would be rapidly stifled. Sound could not travel such distances as are actually observed, unless the compressions are either isothermal or adiabatic, and Laplace's calculation enables us to say that they must be adiabatic.

We shall now use the formula $V = \sqrt{\dfrac{\gamma p}{\rho}}$ to find how the velocity of sound in a gas depends on the nature of the sound waves, and on the pressure, density and temperature of the gas.

106. Velocity almost independent of Amplitude.
We see in the first place that neither the wave-length nor the amplitude of the waves appears in the expression, and therefore to a first approximation waves of all lengths and amplitudes travel with the same velocity. This is true only when the amplitude is small, as is generally the case. It will be remembered that in proving the coefficient of elasticity to be γp we neglected certain small terms. If these terms were retained the coefficient would be a little greater. We have assumed for gases a law similar to Hooke's Law for solids, and it is not strictly correct to make this assumption except when the amplitude is infinitely small. For all ordinary sounds it is sufficiently accurate to take the velocity as independent of the amplitude, but very loud sounds, such as the sound of a cannon, do in fact travel appreciably faster than ordinary sounds.

107. Velocity independent of Pitch. Since the velocity of sound is the same for all wave-lengths and amplitudes provided the amplitude is not very great, we see from the equation $v = n\lambda$ that the wave-length is inversely proportional to the frequency; that is to say, the higher the pitch of a note the less is its wave-length. The lowest audible note has a frequency 30. If we assume the velocity of sound is 1090 ft. per sec. we find from the equation above that the wave-length of the note is 36·3 ft. The highest audible note has a frequency about 20,000, whence its wave-length is $\frac{2}{3}$ inch. A man's ordinary speaking voice has a wave-length of about 8 feet, and a woman's about 4 feet.

108. Velocity independent of Pressure. It is clear that a change in the pressure of the air has no effect on the velocity of sound, for the density changes in the same proportion as the pressure and p/ρ is unchanged. Hence a rise or fall of the barometer makes no change in the velocity, and sound travels with the same velocity at high altitudes as at the sea level.

109. Velocity in different gases. If we compare the velocity in any one gas with that in another of different density, we see that the velocities are inversely proportional to the square roots of the densities, provided γ has the same value for the two gases. Oxygen and hydrogen, for instance, have the same γ, but oxygen is 16 times as dense as hydrogen, and therefore the velocity of sound in hydrogen is 4 times as great as in oxygen. We shall see later, when we treat of organ pipes, that a whistle blown with hydrogen would give a note two octaves higher than the same whistle blown with oxygen at the same temperature, for the pitch of the whistle depends on the number of times the sound can travel the length of the pipe in a second. The change of pitch due to change of density is easily shewn by blowing a whistle first with air and then with coal gas.

Blaikley has pointed out (*Cantor Lecture*, 1904) that, if the earth's atmosphere had consisted of hydrogen instead of air, a Piccolo would have needed to be a yard long, and a contra-bass Saxhorn as long as a cricket pitch, to give notes of the same pitch as at present.

110. Effect of Temperature on Velocity. In order to find the effect of temperature we may write the expression for the velocity in the form $\sqrt{\gamma p v}$. We know from Charles' Law that $pv = kt$, where k is a constant for any one gas, and t is the absolute temperature, and therefore we have $V = \sqrt{\gamma k t}$ or the velocity of sound in a gas is proportional to the square root of the absolute temperature of the gas. If C is the temperature on the Centigrade scale

$$V = \sqrt{\gamma k (C + 273)}$$

and if F is the temperature on the Fahrenheit scale

$$V = \sqrt{\gamma k (F + 459)}.$$

In order to get a numerical estimate of the effect of temperature let us suppose the temperature rises from 60° F. to 61° F., then we have

$$\frac{\text{Velocity at } 60°}{\text{Velocity at } 61°} = \frac{\sqrt{519}}{\sqrt{520}} = \frac{1039}{1040} \text{ nearly.}$$

As the velocity of sound at 60° is not far from 1039 feet per second, it follows that the velocity of sound in air increases by about one foot per second for each Fahrenheit degree rise of temperature at ordinary temperatures.

111. Velocity in Mixtures of Gases.
If we have a mixture of two or more gases, we must use the actual density of the mixture in calculating the velocity of sound. If the γ of the constituents is not the same, the appropriate value to be used must be calculated from the values for the separate constituents*.

The mixture with which we are most usually concerned is moist air, which is merely a mixture of water vapour and air. Water vapour is lighter than air in the ratio of 9 to 14·4; hence the more moisture there is in the air the greater is the velocity of sound.

As an illustration of the use of the formula let us find the velocity of sound in dry air at 20° C., assuming the velocity in dry air at 0° C. to be 332 metres per sec. We have

$$\frac{\text{velocity at } 20°}{\text{velocity at } 0°} = \frac{\sqrt{293}}{\sqrt{273}}.$$

$$\therefore \text{ velocity at } 20° = \sqrt{\frac{293}{273}}\, 332 = 344.$$

Next suppose the air is saturated with moisture at 20°, and that the height of the barometer is 760 mm. The maximum vapour pressure of water vapour at 20° C. is 17·4 mm., hence the partial pressure of the water vapour is 17·4 and that of the air 742·6 mm. If the density of hydrogen be taken as 1, that of air is 14·4, and that of water vapour 9 under the same conditions of temperature and pressure. The density of the moist air will then be in the same units

$$\frac{17·4 \times 9 + 742·6 \times 14·4}{760} \text{ or } 14·28.$$

---

* If P is the pressure and Γ the ratio of the specific heats of the mixture it can be shewn that

$$\frac{P}{\Gamma - 1} = \frac{p_1}{\gamma_1 - 1} + \frac{p_2}{\gamma_2 - 1} + \text{etc.},$$

where p_1, p_2, etc. are the partial pressures, and γ_1, γ_2, etc. the ratios of the specific heats of the constituents of the mixture.

The value of γ for water vapour is about 1·3, whilst that for air is 1·41, but as the amount of water vapour in the air is so small, the effect of this difference is inappreciable, and we may assume that γ for the wet air is the same as for dry air. We may therefore write

$$\frac{\text{velocity of sound in wet air at } 20°}{\text{velocity of sound in dry air at } 20°} = \frac{\sqrt{14·4}}{\sqrt{14·28}}.$$

Or the velocity in wet air at 20°

$$= \sqrt{\frac{14·4}{14·28}}\, 344 = 345·3 \text{ metres per sec.}$$

Thus the effect of saturating the air with moisture at 20° C. is to make sound travel faster by 1·3 metres per sec.

112. Velocity of Sound in Liquids. Longitudinal waves can travel through liquids in the same way as they travel through air, and the expression for their velocity is the same as that for air.

Liquids require very great forces to compress them, or in other words their coefficient of volume elasticity is very great. This large coefficient of elasticity is more than enough to compensate their greater density, and it is found that the velocity of sound in liquids is in general greater than in gases. The velocity of sound in water, for instance, is about 1450 metres per sec., which is nearly 5 times as great as the velocity in air.

113. Velocity of Sound in Solids. We can have longitudinal waves in solids, but the expression for their velocity is not so simple as in the case of gases and liquids. A solid has two fundamental coefficients of elasticity, a volume coefficient and a rigidity coefficient, and both are involved in the expression for the velocity.

When a longitudinal wave travels along a thin rod of metal, there is lateral expansion of the rod at regions of compression, and shrinkage at regions of rarefaction. In this case the coefficient of elasticity concerned is Young's Modulus, the ratio of the force used to stretch a rod of unit section to

VELOCITY OF LONGITUDINAL WAVES

the resulting extension of unit length of the rod. If longitudinal waves travel through a *mass* of the same metal the lateral expansion and shrinkage cannot take place, and the coefficient of elasticity is greater than in the case of a rod. It follows that the waves travel more slowly through a rod than through an extended mass of the same material.

The methods used for measuring the velocity of sound experimentally will be given in Chapter XII.

114. Velocity of Sound in Various Media. The annexed table gives the velocity of sound in several media. The last five must not be taken as more than rough approximations. Different samples of the materials vary greatly in their properties, and different observers have found very different values for the velocity of sound.

Air at 0° C.	332	metres per second
Hydrogen	1270	,, ,, ,,
Carbon Dioxide	257	,, ,, ,,
Water	1450	,, ,, ,,
Iron	5000	,, ,, ,,
Oak	3400	,, ,, ,,
Pine	5200	,, ,, ,,
Glass	5500	,, ,, ,,
Vulcanized indiarubber	43	,, ,, ,,

CHAPTER VI

REFLECTION AND REFRACTION. DOPPLER'S PRINCIPLE

115. Reflection of Spherical Waves. We have already dealt in Chapter IV with the reflection of sound at the closed or open end of a pipe. We must turn now to the more general case of reflection of waves in the open air.

Let A (Fig. 44) be a source of sound from which spherical waves spread out in every direction, and let the circles drawn with A as centre be the sections of those spherical wave-fronts which shew the positions of the surfaces of maximum compression at the moment for which the figure is drawn. The radius of each of these circles increases by about 1100 feet per second, if the air has its ordinary temperature, and the distance between any two consecutive circles measured along a radius is the wave-length of the sound.

Let MN be a rigid wall and O the point where the perpendicular from A meets the wall. At the moment for which the figure is drawn the wave-front 6 touches the wall at O. A little later the part of the wave-front that is not hindered by the wall will have come into the position 7, but the part between C and D will have been reflected towards the left. The part near O reached the wall first and was reflected first, the parts farther from O were reflected a little later, whilst the parts at C and D have just come into a position to be reflected. Consequently the wave-front between C and D will have taken some such shape as the curve CED; that is to say, the wave-front which had the form of the sphere $6O6'$ has now been bulged inwards between C and D, and the bulge is travelling towards the left. As the original wave-front continues to spread outwards from A the bulge will get deeper and wider, and will occupy the successive positions shewn in the figure.

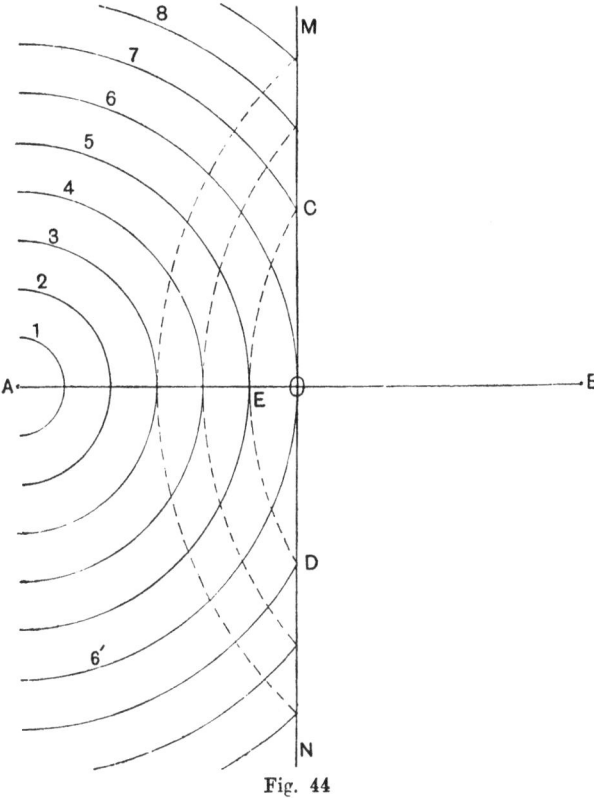

Fig. 44

It can be shewn that the bulges are spheres spreading outwards from B as centre, where B is as far behind the wall as A is in front of it, and their velocity is the same as that of the original wave-fronts.

116. Sound Images. These spheres form the reflected wave-fronts; and we see therefore that the sound after reflection appears to radiate from the point B. Using the analogy of light we may say that B is the sound image of A. The construction for finding B is the same as would be employed if it were required to find the image of A, if A were a source of light, and MN were a mirror.

In Optics it is often more convenient to solve problems by making use of rays instead of wave-fronts, and we may use the same method here. We should then look on A as sending out rays of sound in every direction; any ray such as AC being reflected by the wall in such a direction that the incident and reflected rays make the same angle with the normal to the wall.

If AC and CF make the same angle with the normal the triangle COB formed by continuing FC backwards until it meets AO produced is equal in all respects to the triangle ACO, and therefore OB is equal to AO. The same applies to every ray from A that strikes the wall, and consequently all the rays will after reflection appear to have come from B, if our assumption that the incident and reflected rays make the same angle with the normal is correct. We shall return to this point presently.

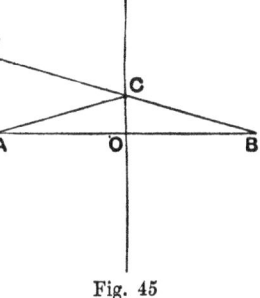

Fig. 45

117. Echoes. We have now the explanation of the simplest kind of echo. Suppose a person situated at A (Fig. 45) makes a sharp sound, such as that produced by clapping the hands, or firing a gun, and there is a listener at F. The sound reaches F by two paths, one of which is the direct line AF, and the other is the broken line ACF. These paths are of different lengths and therefore two sounds will reach F, the first by the path AF, and the second by the path ACF. The interval of time between them will be the time sound takes to travel the difference between the lengths of the two paths. Suppose for instance that AF is half a mile and $AC + CF$ is a mile and a half. The second sound has to travel a mile farther than the first, and so will reach F nearly 5 seconds later.

It is possible to make a rough determination of the velocity of sound by making use of an echo. Stand about 100 yards from a high wall or the side of a house and clap the hands. An echo will be heard, and if by some means the interval between the clap and the arrival of the echo is measured, we shall know the time taken by the sound to travel to the wall and back along a line perpendicular to the wall. The velocity could not be measured very accurately in this way, as in the case given the interval between the clap and the arrival of the echo would be only a little over half a second, which is too short to measure without special appliances. Moreover, in the open air most of the sound spreads out sideways and is lost, so that the echo is faint unless the sound is loud. The experiment can be carried out more easily in a tunnel with a closed end or in a cloister. Stand near one end of the cloister and clap the hands at regular intervals, timing the intervals to one second by means of a watch, or a metronome which ticks seconds. Now move slowly away from the wall, and presently an echo will be heard following each of the claps. Continue to move from the wall until the echoes fall exactly halfway between the claps, which can be judged with considerable accuracy by the ear, and measure the distance from the wall when this happens. It is clear that the sound now takes half a second to travel to the wall and back, and if, for instance, the distance of the observer from the wall is found to be 275 ft., the velocity of sound is 1100 ft. per second.

118. Reflection of Sound by Spherical Mirrors.
We have assumed that sound is reflected according to the same laws as hold for light, namely:—

(1) The incident ray, the reflected ray, and the normal to the reflecting surface are in the same plane.

(2) The incident ray and the reflected ray make equal angles with the normal.

It is shewn in books on Optics that, if these laws are true, a beam of parallel rays falling normally on a spherical mirror will come together at the principal focus of the mirror after reflection; the principal focus being the centre of a radius of the mirror drawn in the direction in which the incident beam travels. Conversely, if a source of light is placed at the principal focus, the rays diverging from it so as to strike the mirror will be reflected as a parallel beam.

A similar result is obtained experimentally when a source of sound is used instead of a source of light, and the experiment may be taken as a verification of the two laws stated above.

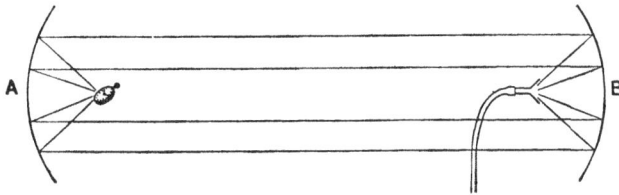

Fig. 46

Place two large spherical mirrors facing each other, and 20 to 30 feet apart. At the focus of one of the mirrors place a watch that ticks fairly loudly. The sound waves which strike the mirror A will be reflected in a parallel beam, and will therefore travel without loss of intensity to the mirror B, where they will be reflected again and converge to the focus. There will consequently be a concentration of the sound at the focus of the second mirror, and the ticking of the watch will be heard clearly if a small funnel connected with a rubber tube leading to the ear be placed at the focus, but will be inaudible if the funnel is moved a little distance away. The mouth of the funnel must be pointed towards the mirror B, for the sound is heard by means of rays that have been reflected from the two mirrors, and not by means of rays that come directly from the watch to the funnel. These latter rays are divergent, and are too weak to affect the ear.

The experiment can be shewn in a striking manner to a number of persons at once by using a sensitive flame instead of a funnel and tube. To make a sensitive flame draw out a piece of glass tubing to a fine point, and connect it to a gas bag of coal gas. Light the gas issuing from the narrow tube, and gradually increase the weights on the gas bag until the flame is on the point of roaring. The flame will now be steady and a foot or more in height according to the size of the jet, but if any sound of high pitch is made in its neighbourhood, it will become much shorter and will roar. The rattle of a bunch of keys or a hiss will make it drop at once, and if one talks in its presence, it will drop whenever the letter S is pronounced. If such a sensitive flame is placed with the point of the glass tube at the focus of the mirror B, the flame will dance in time with the ticks of the watch, but the dancing will cease if the jet is moved a little to one side, for it is only at the focus that the sound is strong enough to affect the flame.

119. Concentration of Sound by the Walls of a Room. One occasionally meets with an instance of this concentration of sound by curved surfaces in a large room. If the opposite ends of the room are curved, a person speaking at the focus of one end is heard distinctly by a person standing at the focus of the other end. As the curved wall is generally part of a cylinder and not part of a sphere, the focus is not a point but is spread out into a line. The concentration of the sound is therefore not so great as it is when the reflecting surfaces are spherical.

It is not essential that the source of sound should be at the principal focus of the first mirror. It must be on or near the common axis of the two mirrors, but there will be a concentration of the sound at some point, wherever the source is situated along the axis. This follows from the analogy of the corresponding optical phenomenon.

It is not even necessary that there should be two mirrors. A light placed anywhere on or near the axis of a spherical mirror, and farther from the mirror than the principal focus, will have a real image somewhere, the light and its image being situated at a pair of conjugate foci; and the same is true of sound. There are stories of churches where one can hear at certain points what is said in distant confessional boxes. If the stories are true, it is probable that the sound of the voices is collected by some curved surface and concentrated at a distant point.

120. Whispering Galleries.

The well-known Whispering Gallery of St Paul's owes its peculiar acoustical properties to the reflection of sound by the walls. The gallery is in the form of a circle running round the base of the inside of the dome.

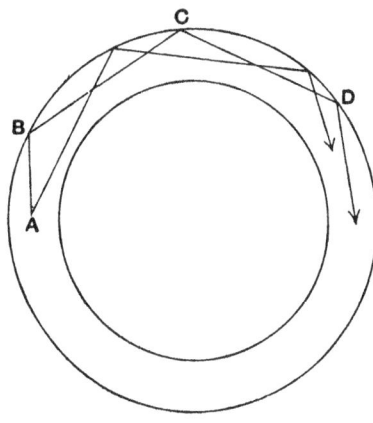

Fig. 47

If a person at any point A puts his head close to the wall and whispers, the whisper can be heard at any part of the gallery, if the listener also places his head near the wall. This is not a case of the concentration of the sound at a focus, for the whisper can be heard almost equally well at any part of the gallery, and the listener can easily convince himself that the sound creeps round the dome in a thin sheet close to the wall.

If the dome were not present, the sound from A would spread out equally in all directions, and would fall off in intensity inversely as the square of the distance from A. The effect of the dome however is to prevent much of the spreading.

A ray AB, for instance, which starts in a direction that does not make a large angle with the tangent to the wall near A, will take the path $ABCD$, etc., and so remain near the wall, and many other rays follow similar paths, so that the sound remains strong enough to be audible at any point of the circumference.

121. Musical Echo from a Row of Palings. If a sharp sound is made near the end of a row of palings, the echo sometimes takes the form of a musical note. Each of the palings reflects the sound in turn, and the further they are from the listener, the longer the sound takes to return, so that the result is that we have a series of echoes following each other in rapid succession, and if the palings are sufficiently close together, the echoes will blend into a musical note.

122. Experimental demonstration of Stationary Vibrations in Air. We saw in Chapter IV that, when a train of waves is reflected from a solid obstacle, the direct and reflected trains combine to give a series of stationary vibrations with nodes half a wave-length apart, the reflecting surface being at a node. This phenomenon can be conveniently investigated experimentally with the aid of a sensitive flame.

A note of high pitch is needed for the experiment, for the flame is not affected by notes of medium or low pitch. A whistle that gives a note suitable for the purpose can be made in the following way. Take a brass tube six inches long and an inch in diameter, and close one end with a very thin plate pierced with a hole $\frac{1}{20}$ in. in diameter. Take a second tube eight to ten inches long, and of such a diameter that it slides tightly over the shorter tube, and fix at its centre a thin plate also pierced with a hole $\frac{1}{20}$ in. in diameter. Slide the wider tube over the closed end of the narrower tube, until the two pierced plates are half an inch or less from each other, and connect the open end of the narrower tube with a gas bag full of air. The whistle will be found to give a high note which has a powerful effect on the sensitive flame. The pressure on the gas bags supplying the whistle and flame should be regulated so as to give the best effect, and should be kept constant as the pitch of the note depends on the pressure.

Place the whistle with its mouth several feet from a vertical board, and bring the sensitive flame between the whistle and the board. As the flame is moved along the normal from the whistle to the board, there will be found a series of equidistant points at which it is undisturbed. These points are the nodes of the stationary waves, and the distance between any two consecutive nodes is half a wave-length. The experiment gives us the means of finding the pitch of the whistle, for we know the velocity of sound, and can find the wave-length by doubling the distance between two consecutive nodes, and so we can calculate the frequency n from the equation $v = n\lambda$.

Fig. 48

The wave-fronts here are spheres with the mouth of the whistle as centre, and therefore the circumstances are not the same as in the case where stationary waves are formed in a tube as explained in Chapter IV. Since the flame is moved along that radius of the spheres which is normal to the board, the tangents to the wave-fronts at points on the path of the flame are all parallel to the board. The only rays with which we are concerned lie very close to the normal, the oblique rays being reflected away from the flame, and we may therefore regard the wave-fronts as being small parallel planes. The nodes will not all be equally well marked, for the sound falls off in intensity as it gets farther from the whistle. The farther a node is from the board, the greater will be the difference between the intensities of the direct and the reflected waves, and the less well marked will be the node. Measurements should therefore be made as close to the board as possible. If accurate measurements are desired some means must be adopted for preventing the reflection of waves from the table on which the apparatus stands, for these waves alter the positions of the nodes. A sheet of cotton wool or felt will destroy the greater part of the reflection.

123. Refraction of Sound. We have a further analogy with light in the refraction of sound. When plane sound-waves cross the boundary separating two media in which their velocity is different they are deflected, their direction of propagation making a smaller angle with the normal to the separating surface in the medium in which they have the smaller velocity.

The reason for the refraction need not be given here as it is the same as for light, and can be found in any treatise on optics. The laws of the refraction of sound too are the same as for light. The incident ray, the normal to the refracting surface, and the refracted ray are in the same plane; and the sine of the angle between the incident ray and the normal bears a constant ratio to the sine of the angle between the refracted ray and the normal, this ratio being equal to the ratio of the velocities of sound in the two media.

124. Total Internal Reflection of Sound. The case of Total Internal Reflection has an interesting application in acoustics.

Let AB be the surface separating two media of which the lower is that in which the sound travels more slowly, and let CO be a ray of sound incident at O. By the second Law of Refraction we have

$$\frac{\sin COE}{\sin DOF} = \frac{v_1}{v_2},$$

where v_1 and v_2 are the velocities of sound in the two media. If CO just grazes the surface,

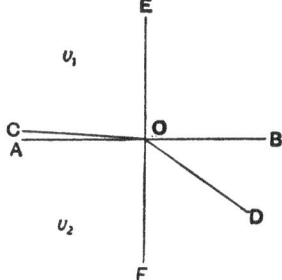

Fig. 49

$COE = 90°$ and $\sin COE = 1$, whence $\sin DOF = v_2/v_1$. Now the path of any ray, whether of light or sound, is reversible, and therefore if DO is the incident ray, the refracted ray will graze the surface on emergence into the upper medium. If

100   EFFECT OF WIND   [CH. VI

the incident ray makes any greater angle than FOD with the normal it cannot emerge at all into the upper medium, for if it did emerge it would violate the second law. It is then totally reflected at the separating surface, and remains in the lower medium. The critical angle beyond which total reflection takes place is the angle which has a sine equal to v_2/v_1. Let us suppose the lower medium is air and the upper is pine wood. The velocities of sound in the two media are 332 and 5200 metres per second respectively. Consequently the sine of the critical angle is $\frac{332}{5200}$, and the angle is about $3\frac{1}{2}°$. Thus we see that unless the sound strikes the surface very nearly normally none of it will get into the wood, and conversely, if a source of sound is inside a mass of wood, the rays which emerge into the air will nowhere make an angle of more than $3\frac{1}{2}°$ with the normal at the point where they emerge.

We see too why sound loses so little in intensity when it travels along a pipe, for it cannot get out unless it strikes the walls nearly normally.

125. Effect of Wind on Sound. It is well known that sounds travelling with the wind are heard better than those travelling against it. This is due to a phenomenon analogous to refraction.

Let us suppose that both sound and wind are travelling from left to right, and A (Fig. 50) represents a wave-front at any moment. The wind travels more slowly near the ground than higher up, and the upper part of the wave-front is helped forward more than the lower part. Consequently, as the wave-front moves on, it will tilt forward into the positions B, C, D, etc. Now the sound always travels in a direction at right angles to the wave-front, and therefore any ray drawn at right angles to all the wave-fronts takes a path which curves downwards towards the ground like the curved line in Fig. 50.

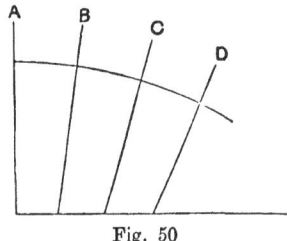

Fig. 50

A person when speaking sends out rays of sound in every direction. If the air is still the rays are straight, and so

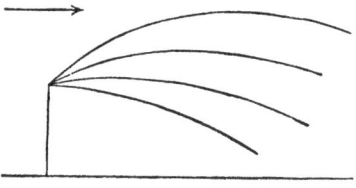

Fig. 51

most of them go above the listener and are lost. If however the wind blows from the speaker to the listener, the rays that travel with the wind curve downwards and many that would otherwise be lost are able to reach the listener. The effect is specially marked when there are obstacles between the speaker and the listener, for the curvature of the rays enables the sound to rise over the obstacle and come down on the other side.

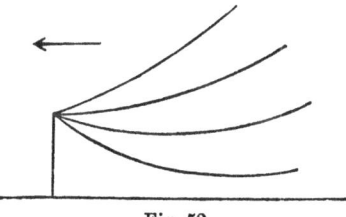

Fig. 52

When the sound travels against the wind the effect is reversed. The upper part of a wave-front is retarded more than the lower part. The wave-front therefore tilts backwards, and the sound rays curve upwards as in Fig. 52. The result of this is that at some little distance from the speaker there are no sound rays remaining near the ground, and the sound is inaudible.

126. Effect of Varying Temperature of the Air.
A similar effect is observed when there is a gradual change in the temperature of the air from the ground upwards.

Suppose, for instance, the temperature rises as we go upwards. The warmer the air, the greater is the velocity of sound in it, and therefore the result will be the same as when sound travels with the wind. The upper parts of the wave-fronts will travel faster than the lower parts, and the sound rays will curve downwards. Conversely if the temperature of the air gets lower from the ground upwards, the sound rays will be deflected upwards and will be lost in the upper air.

127. Sound Shadows. There is one marked difference between light and sound which merits a little explanation, though the complete explanation is beyond the scope of this book. Sound and light both consist of wave motion, and we have seen that many of the phenomena of light have their counterpart in the case of sound. In the matter of the casting of shadows, however, there appears at first sight to be a difference. If an opaque obstacle is placed in the path of a beam of light, no light in general gets round to the back of the obstacle, but we have a sharply defined shadow. If an obstacle is placed in the path of a beam of sound, this effect is generally almost absent. The sound appears to be able to get round corners quite freely, and is heard almost as well behind the obstacle as in front of it.

The apparent anomaly arises from the difference in the ratio of the wave-length to the dimensions of the obstacle in the two cases. It can be shewn that a sharp shadow is formed only when the obstacle is large compared with the wave-length, whether the waves be those of light or sound.

The wave-length of light varies between $\frac{1}{35000}$ in. and $\frac{1}{50000}$ in. according to its colour. Any ordinary obstacle is much greater than this in diameter, and therefore light in general casts sharp shadows. If however the obstacle is very small, such as a fine wire, it is found that some light gets round to the back, and the shadow is not sharply bounded.

The wave-length of a man's ordinary speaking voice is 8 to 10 feet, and that of a woman's voice is 4 to 5 feet. Now the obstacle must be at least 50 wave-lengths in diameter to cast a shadow and even then the shadow would be badly defined, so that we cannot expect ordinary objects such as walls or houses to have much effect in cutting off the sound of the human voice. Practically the only weakening effect such objects have arises from their compelling the sound to take a longer path from a point in front to a point behind them.

When the pitch of the note is high and the obstacle large, the sound shadow may be very marked. The writer has met with a striking instance of this on Pilling Moss in North Lancashire. In the Spring the sea-gulls resort in large numbers to the Moss to lay their eggs, and when the young birds are able to fly, the air is filled with their shrill screams. There is a road at a little distance from the nests, and by the side of the road there is sometimes a row of stacks of peat. The length of one of these stacks is many times as great as the wave-length of the screams of the birds, and consequently a good sound shadow is formed. As one walks along the road the alternations of sound and silence are very marked. Opposite the gap between two stacks the sound is unpleasantly loud; opposite the stack itself there is almost complete silence, and the change from sound to silence is quite sudden.

A similar effect can be obtained in the laboratory, though it is less marked. If a card two feet square is held between the ear and the high pitched whistle described earlier in this chapter, the sound is perceptibly weakened. If a medium pitched tuning-fork is used instead of the whistle, the interposition of the card has little or no effect in weakening the sound.

It should be mentioned here in anticipation of what will be said in the chapter on Quality that sounds are generally composite, consisting of tones of various pitches, and the quality of a sound depends on the number, pitch, and intensity of the tones that are present. Some of these additional tones may be of high pitch, so that a given obstacle may be able to cast a sound shadow for the higher constituents, whilst having little effect on the lower constituents. Consequently at the back of the obstacle such a sound will be found to have a different quality, from the weakening of the higher constituents relatively to the lower ones.

128. Doppler's Principle. When a source of sound is moving to or from a stationary listener, the pitch of the note heard by the listener is not the same as when the source is stationary. Similarly a listener moving to or from a stationary source of sound hears a note of different pitch from that which he would hear if he were not moving.

This may often be observed at a railway station. If a passing train whistles as it goes through the station, the pitch of the note given by the whistle falls just as the train passes a listener on the platform. When the train is approaching, the pitch is higher than it would be if both train and listener were at rest, and when the train is receding, the pitch is lower. If the train is travelling at 40 miles per hour, the fall of pitch is nearly a tone.

The effect can be shewn in the laboratory by holding in the hand a vibrating tuning fork mounted on a resonance box, and swinging it rapidly in a circle. The person who is swinging the fork will hear no change of pitch, because the fork is always at approximately the same distance from his ear, but any other person, who places himself in the plane in which the fork swings, will hear that the note is higher when the fork is moving towards him than it is when it is moving from him.

We may say in general terms that when the source and the listener are approaching each other the pitch of the note is raised; when they are getting farther apart the pitch is lowered.

It is not, however, a matter of indifference whether it is the source or the listener that moves. If the whistle and the observer are approaching each other with a given velocity, the observer will hear a note of rather higher pitch if he is at rest and the whistle is moving towards him than he would hear if the whistle were at rest and he were moving towards it.

The explanation of these phenomena given in §§ 129–132 is known as *Doppler's Principle*.

129. Source in motion and Listener at rest.
Let us suppose first that the listener L (Fig. 53) is at rest, and the source S is moving towards him.

The source is producing waves at a definite rate n per second, n being the frequency of the note that would be heard if source and listener were at rest. These waves travel towards L with a velocity v, and the source follows them with a velocity V. At the end of one second S will have reached S', where $SS' = V$, and the wave sent out at the

```
S        S'                    O                      L
```
Fig. 53

beginning of the second will be at O, where $SO = v$. Now the source has given out n waves during the second, and these n waves must all be between S' and O, and therefore the length of each wave is $\dfrac{S'O}{n}$ or $\dfrac{v-V}{n}$. If the source had been at rest, the n waves would have been distributed over SO, and the wave-length would have been $\dfrac{v}{n}$, so that the effect of the motion of the source is to shorten the wave-length in the ratio $v - V$ to v. Now the waves travel with the same velocity whatever their length may be, and so the listener receives in one second as many waves as there are in a length v, that is $\dfrac{v}{v-V} n$. If then we write n' for the frequency of the note heard, we have $n' = \dfrac{v}{v-V} n$. It should be noted that the waves are actually shorter than they would be if the source were at rest.

130. Listener in motion and Source at rest.
Next let us suppose that the source is at rest, but the listener is moving towards it with velocity V.

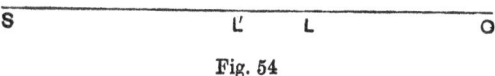

Fig. 54

Let the source be at rest at S, and at the moment we are considering let the listener be at L. The space between S and L is filled with waves of length $\lambda = \dfrac{v}{n}$. These waves travel with velocity v, and therefore the particular wave that is now at L

will one second later be at O, where $LO = v$. All the waves between L and O have passed through L during the second, but meanwhile the listener has moved a distance V to L', and so has received not only the waves now between O and L, but also those now between L and L'; that is he has received in one second all the waves that at the end of the second are included in a length $V + v$. The wave-length is not affected by the motion of the listener, and therefore there are n waves in LO and $\dfrac{V+v}{v} n$ in $L'O$. Consequently if n'' is the frequency of the note heard by the listener, $n'' = \dfrac{v + V}{v} n.$

It will be seen that this is not the same formula as was obtained for the case where the source is in motion, and the method of deducing the formulae shews that we ought not to expect them to be the same, for in the one case the change of pitch is due to a real change of the wave-length, and in the other case it is due to a change in the rate at which the waves pass the observer, though there is no change of wave-length.

131. Source and Listener in motion. The two formulae can be combined so as to give an expression for the pitch of the note heard when both source and listener are moving. Let V_1 be the velocity of the source and V_2 the velocity of the listener both being reckoned positive when the velocity is from left to right.

Then if the listener is at rest we have as before

$$n' = \frac{v}{v - V_1} \cdot n.$$

If now the listener be moving we have

$$n'' = \frac{v - V_2}{v} n',$$

V_2 having its sign changed, because we treated V_2 as positive in the investigation above, when the listener was moving towards the left.

Substituting in the second of these equations the value of n' given by the first we find

$$n'' = \frac{v - V_2}{v - V_1} n \quad \ldots\ldots\ldots\ldots\ldots\ldots\ldots(1).$$

It follows from this that if $V_1 = V_2$ then $n'' = n$, or there is no change of pitch if the source and listener move with equal velocities in the same direction, and so remain the same distance apart. When a person hears the whistle of the engine of the train in which he is sitting, the pitch of the note is the same whether the train is moving or not.

132. Effect of Wind on the pitch of Sound. The formula (1) assumes that the air is still. If a wind is blowing with velocity w in the direction in which the sound travels, we must write $v + w$ in place of v in the formula, and so obtain

$$n'' = \frac{v + w - V_2}{v + w - V_1} n \ \ldots\ldots\ldots\ldots\ldots\ldots(2).$$

If then $V_1 = V_2$ or the source and the observer remain the same distance apart the wind has no effect on the pitch of the note that is heard. If the wind blows from the observer to the source, it is true that the waves will be shortened, just as though there were no wind, and the source were moving towards the observer, but the effect of the wind is to reduce the velocity of sound relatively to the ground and the observer, and this reduction in velocity exactly compensates the effect of the reduction in wave-length on the pitch. It is plain without going into details that wind cannot have any effect in this case, for there is no difference in principle between the case where the source and listener are at rest and the wind blows with velocity V, and the case where the source and listener are moving in the same direction with velocity V, and the air is at rest.

If the source and observer are moving with different velocities, the pitch of the note heard is not independent of the velocity of the wind, as is easily seen by formula (2), for $\dfrac{v + w - V_2}{v + w - V_1}$ is not equal to $\dfrac{v + w' - V_2}{v + w' - V_1}$ unless $V_2 = V_1$.

133. Doppler's Principle applied to Light. A similar phenomenon is observed in the case of light. If a source of light is moving rapidly towards the observer, the wave-length of the light is shortened, and vice versa. The spectrum of a star generally contains bright or dark lines. If the star is approaching the earth, all the lines are displaced a little in the direction corresponding to shorter wave-lengths, that is towards the blue end of the spectrum. If the star is receding, the lines are displaced towards the red end.

By measuring the amount of displacement the velocity of approach or recession can be determined. This method of finding the motion of stars in the line of sight is largely used in astronomical investigations, and has proved of great value.

CHAPTER VII

INTERFERENCE, BEATS, COMBINATION TONES

134. Meaning of the term Interference. In the present chapter we shall discuss some phenomena that result from the superposition of wave trains on each other. When the result of the superposition is to give a distribution of energy very different from that due to the separate trains, the phenomena are generally classed under the name *Interference*.

Before passing to experimental methods it will be useful to find from theoretical considerations what is the result when two harmonic trains of plane waves of the same wave-length are superposed. We shall take the three cases where (1) the waves travel in the same direction in the same straight line, (2) they travel in opposite directions in the same straight line, and (3) they travel in two directions meeting at any angle. We may call these Coincident Trains, Opposite Trains and Oblique Trains respectively.

135. Superposition of Coincident Trains of Waves. In the case of Coincident Trains we have the same difference of phase at every point of the trains and at every moment. Whatever may be the distance between two adjoining crests—one taken from each train—that distance is the same everywhere. If we compound two such trains in the usual way we get merely a third harmonic train of the same wave-length as the constituents, and of an amplitude which depends on the difference of phase of the constituents. The student should find no difficulty in proving this graphically.

*136. Equation of the Resultant Train. We can prove analytically that two harmonic trains of the same wave-length and in the same straight line must always give a third harmonic train.

We saw that we can represent one of the trains by the equation

$$y = a \sin \frac{2\pi}{\lambda}(vt - x),$$

where y is the displacement at time t and at a distance x from the origin, λ is the wave-length, v the velocity of the waves, and a the amplitude.

The second constituent differs from this only in amplitude and phase, and so can be written

$$y' = b \sin \frac{2\pi}{\lambda}(vt - x + a).$$

These when compounded give the wave

$$Y = y + y' = a \sin \frac{2\pi}{\lambda}(vt - x) + b \sin \frac{2\pi}{\lambda}(vt - x + a)$$

which can be written in the form

$$Y = c \sin \frac{2\pi}{\lambda}(vt - x + \beta)$$

if $$c \cos \beta = a + b \cos a$$

and $$c \sin \beta = b \sin a$$

or $$\tan \beta = \frac{b \sin a}{a + b \cos a}$$

and $$c^2 = a^2 + b^2 + 2ab \cos a.$$

137. Distribution of energy in the Resultant Train.

Two special relations of phase should be mentioned. If the constituents are in the same phase at any point, they are so everywhere. If they have the same amplitude they compound into a wave train with double the amplitude and four times the energy of either. If they are in opposite phase and have the same amplitude they neutralize each other everywhere, and there is no displacement, velocity or condensation anywhere.

It is evident from the doctrine of the Conservation of

Energy that this cannot be a complete statement of any case that could actually arise, for we cannot either quadruple or annihilate a stream of energy by adding an equal stream to it. We shall return to this point presently.

It is to be remembered that we are dealing only with plane waves, which do not spread out laterally. If they spread out laterally they will diminish in amplitude as they get farther from their source, and this will modify some of our conclusions. If two spherical sets of harmonic waves of the same wave-length spread out from the same centre, they will compound into a third spherical set, whose intensity will fall off inversely as the square of the distance from the source.

138. Superposition of Opposite Trains of Waves.
We dealt with the superposition of two opposite trains of air waves in Chapter IV, and saw that they compound into a set of stationary waves. At a node there is maximum change of pressure but no motion, whilst at an antinode there is maximum motion but no change of pressure.

139. Effect of Stationary Vibrations on the Ear.
We have not yet enquired how such stationary waves affect the ear. If the ear is moved along a train of stationary waves, will it hear sound everywhere ? If it does not hear the sound equally loud everywhere, will the maximum be at a node or at an antinode ? The experiment could not be made in this crude way, as the head is so large as to interfere with the formation of the waves, but the exploration can be made with a rubber tube, one end of which is placed in the ear and the other moved along the stationary waves. It is found that the loudest sound is heard when the end of the tube is at a node, and nothing is heard when it is at an antinode. The tube merely forms a prolongation of the natural passage leading to the drum of the ear, and when its open end is at a node, the compressions and rarefactions at that point cause waves to run down the tube to the ear. When the open end is at an antinode, the air merely flows to and fro across it, and no sound is heard. Thus when we use the ear as a detector, we may say that there is maximum sound at a node, and no sound at an antinode. It is the variations of pressure that cause the

drum of the ear to vibrate, and not motion of the air without changes of pressure.

140. Superposition of Oblique Trains of Waves.

Let us next consider the superposition of two oblique trains travelling in the directions AO and BO (Fig. 55), and crossing at O; the trains having as before the same wave-length.

Since the two trains have the same frequency, their difference of phase will always be the same at O, but the trains cannot neutralize each other in all respects, whatever that difference of phase may be.

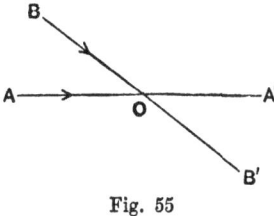

Fig. 55

Let us suppose that they are in opposite phase. We mean by this that, when the particle at O would be moving through its equilibrium position in the direction AA' in which the wave travels if the train AA' only were passing, it would be moving through its equilibrium position in the direction $B'B$ if the train BB' only were passing. These two velocities cannot neutralize each other, for they are not in the same direction, and the same is true of the displacements at any moment. It is clear that the particle at O must describe one of the forms of Lissajous' Figures for the vibration ratio 1:1. If the phases of the constituents are the same, and the amplitudes are the same, the particle will vibrate in a straight line bisecting the angle AOB. If the phases are opposite, it will vibrate in a straight line bisecting AOB', and with any other relation between the phases it will describe an ellipse with O as centre. In Chapter II we dealt only with the figures formed when the directions of vibration are at right angles, but the results would be similar if the directions of vibration made any other angle with each other. Each of the figures would be sheared so as to bring the sides of the enclosing rectangle to the proper angle with each other.

We see then that there will always be velocity and displacement at O, whatever is the difference of phase between the trains, because velocity and displacement have direction as well as magnitude. Compression however has no direction, but only magnitude, and it is possible for a compression due to the train AO to neutralize a rarefaction due to the train BO at the point O. Suppose that the two trains have equal amplitudes and opposite phases. At a certain moment O would be moving through its equilibrium position in the direction AO under the influence of the train AA', and in the direction OB under the influence of the train BB'. It would therefore be in the centre of a region of condensation of the train AA' and in the centre of a region of rarefaction of the train BB'. The result would be no condensation, and this would be the case not only at the moment we have considered, but also at every other moment. There would never be any condensation or variation of pressure, and as regards the effect on the ear O would be a point of silence. If the amplitudes were not equal, there would be minimum sound, though not complete silence, when the phases were opposite.

141. Superposition of two sets of Spherical Waves. Let us now take a more general case and suppose we have two sources of sound of the same frequency and a few wave-lengths apart. For the sake of simplicity let us suppose they are in the same phase. In the space round these sources we shall find regions in circumstances corresponding to the three cases we have discussed.

Let O, O' (Fig. 56) be the two sources. Each source separately would send out waves with spherical wave-fronts. Round the source O draw a series of circles in full lines with radii increasing by λ at each step. In the figure the circles are drawn with radii $\dfrac{\lambda}{2}$, $\dfrac{3\lambda}{2}$, $\dfrac{5\lambda}{2}$, etc. These will shew the positions at some moment of the surfaces of maximum compression due to the waves spreading out from O. Now draw a series of dotted circles half way between those of the first set. These will shew the positions of the surfaces of maximum rarefaction at the same moment. Draw a similar set of full and dotted circles round O'.

Since every point on a full circle is a point of maximum compression for the source round which the circle is drawn, a

114   INTERFERENCE   [CH. VII

point where two such circles cross will be a point of maximum compression, when the two sources send out their waves together. The actual compression will be the sum of those due to the two sources separately. At such a point two rays of sound, one from each source, cross each other in the same phase, and it is easily seen that the difference of the distances of the point from the two sources must be some whole number of wave-lengths.

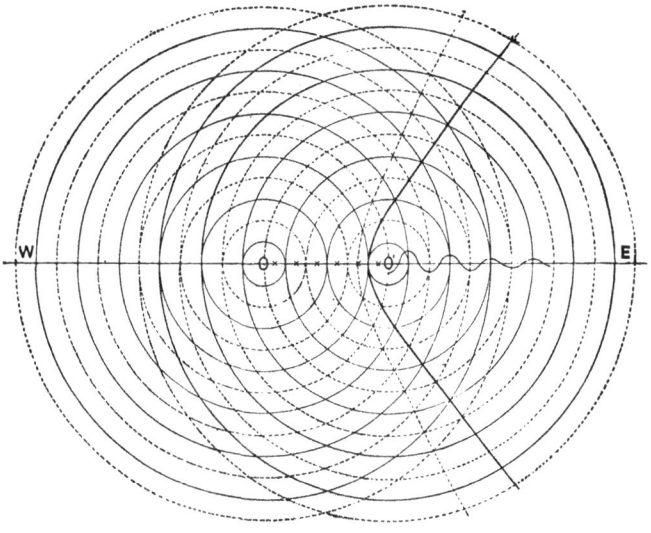

Fig. 56

A point where a full and a dotted circle cross is a point where a compression from one source coincides with a rarefaction from the other, and is therefore a point where the two trains neutralize each other to some extent. The difference between the distances of such a point from the two sources is an odd multiple of half a wave-length. The variation of pressure is not in general completely annihilated, for even if the sources have the same amplitude, the distance of the point from the two sources is not generally the same, and therefore the amplitude of the trains is not the same at that point.

If a curve is drawn through the points where the difference of the distances from the sources is some definite odd multiple of half a wave-length, we get a curve such as the dotted hyperbolic curve in the figure, which is drawn for a difference $\frac{3}{2}\lambda$. This is a locus of points where the compression is a minimum at the moment we are considering. Other similar curves can be drawn for differences $\frac{\lambda}{2}$, $\frac{5\lambda}{2}$, $\frac{7\lambda}{2}$, etc. Moreover these curves are lines of minimum change of pressure at *every* moment; for they are the loci of points where the two separate trains are always in opposite phase.

Along the line $O'E$ we have two trains travelling in the same direction, and therefore compounding into a single progressive harmonic train. If the distance between the sources is an exact number of wave-lengths, the two trains will be in the same phase, and so will give a train with an amplitude at every point equal to the sum of the amplitudes of the constituents. If the distance between the sources is an odd multiple of half a wave-length, the two trains will be in opposite phase, and the amplitude at each point will be the difference of those due to the two constituents.

As the constituent trains diminish in amplitude with increasing distance from the source, the resultant train must do so also, and the intensity of the sound will therefore fall off as we pass outwards along the hyperbolic curve or along $O'E$.

Between O and O' we have two trains of waves travelling in opposite directions, and therefore compounding into a series of stationary waves.

142. Distribution of energy round two Sources.

We see then that the distribution of energy round two sources is very different from that round only one. From one source the energy spreads out uniformly in every direction. From two sources it is concentrated along certain lines. The full hyperbolic curve of Fig. 56 is a line of maximum flow of energy, the dotted curve is a line of minimum flow.

In spite of this redistribution of the energy the total amount flowing outwards is the sum of the outputs of the two sources. The interference neither annihilates nor creates energy on the whole.

143. Exceptional cases where the Sources influence each other. The statement above generally applies when we have interference, but there are exceptional cases where the addition of a second source diminishes the total energy. In this case the proximity of the two sources causes both to give out less energy than they would if far apart. We have an instance of this when two similar organ pipes are mounted close together on the same wind chest. If the two are blown together, the sound produced is everywhere much less than would be produced by one pipe alone. The pipes at once fall into opposite phases, and very little energy is given out by either. The explanation of this action is beyond the limits of this book.

It is possible to have the converse case. If two sources of sound equal in period and amplitude are placed very close together, and compelled to vibrate in the same phase, the amplitude will be double and the energy four times that due to one of the sources. In this case the presence of each source compels the other to do double the work it would do if it were alone.

144. Interference by use of a Branched Tube. A convenient form of apparatus for shewing interference is illustrated in Fig. 57. AC is a tube open at A and C and divided into two branches B and D, one of the branches D having a telescopic slide by which its length can be varied.

A tuning-fork is placed before the opening A, and a rubber tube connected with C carries the sound to the ear. If the slide is adjusted so as to make the two branches of the same length, the two trains of waves that have come by way of B and D respectively will arrive at E in the same phase. They will therefore add their effects and the sound will pass on to the ear as it would have done if the tube had not been divided.

Now draw out the slide gradually, making the branch

D longer than B. The waves arriving at E by way of D will fall more and more behind those coming by B, and when the difference in path between the two branches is half a wavelength, the two trains will neutralize each other at E, and no sound will be heard at C. If part of the branch B consists of a piece of indiarubber tube, it will be found that pinching the tube so as to stop the stream coming by B restores the sound, thus proving that the silence is due to the superposition of the two trains of waves at E.

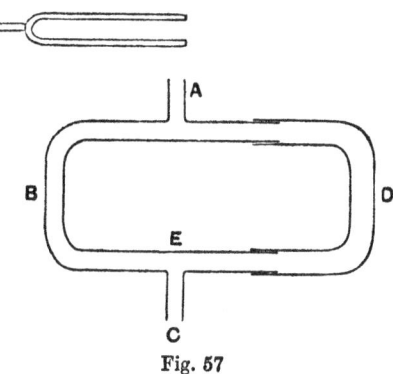

Fig. 57

It may be asked what has become of the energy in this case. The answer is that it has been reflected at E and has returned to the open end A. As the two trains are in opposite phase at E, there is no variation of pressure there. It is because we have approximately this condition at the open end of a pipe that most of the sound is reflected back from an open end with reversal of phase. E then behaves as an open end and the sound is reflected back to A.

145. Interference near a Tuning-fork. Interference may be illustrated in a simple manner by the use of a tuning-fork, though the complete explanation of the phenomenon is less simple than in the case of the divided tube.

Hold a vibrating fork near the ear, and twist it round by turning the shank between the finger and thumb. The sound will be heard to rise and fall in intensity four times in each revolution.

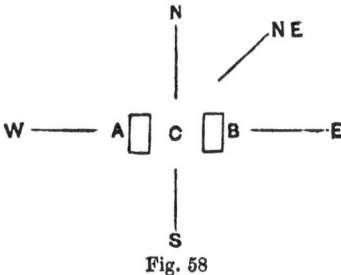

Fig. 58

Let A and B (Fig. 58) represent the free ends of the prongs of the fork. The prongs move inwards and outwards together. When they are moving outwards compressions are produced on the faces A and B, and these compressions travel outwards with maximum amplitude in the directions E and W, and with minimum amplitude in the directions N and S. At the same moment a rarefaction is produced between the prongs at C, and this rarefaction travels outwards with maximum amplitude in the directions N and S and minimum amplitude in the directions E and W. There will then be four directions such as NE where the compression and rarefaction have the same amplitude, and so neutralize each other.

It follows that if the ear moves round the fork, or, what comes to the same thing, the fork is turned before the ear, the sound will be heard plainly in the directions N, E, S, W, but will be inaudible in the direction NE and the three corresponding directions.

146. Seebeck's Interference Tube. Seebeck's Tube gives a good illustration of interference, and also provides the means of measuring the velocity of sound in air with fair accuracy.

It consists of a glass or metal tube with a short side tube A near one end, and a sliding piston B by which the length of the main tube can be varied.

A tuning-fork is placed before the open end C, and a rubber tube leads from A to the ear. The waves from the fork reach A by two paths. Part go from C to D, and then pass out by A. The rest go from C to B, are reflected there, return to D, and pass out by A. Thus one path is longer than the other by twice DB.

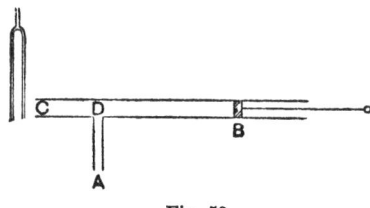

Fig. 59

Move B back and forwards until a position is found for which the sound heard at A is a minimum. The two trains of waves are then in opposite phase at D, and twice DB must be some odd multiple of half a wave-length. It is easy to secure that twice DB is one half wave-length by beginning with the piston near D, and drawing it backwards until the first minimum is reached. If the tube is long enough to give several minima of sound, it is clear that the corresponding positions of the piston will be half a wave-length apart.

If B is in the position for the first minimum we have

$$DB = \tfrac{1}{4}\lambda,$$

and if n, the vibration number of the fork, is known, we have

$$v = n\lambda = 4nDB.$$

This method of measuring the velocity of sound in air has the advantage that only a small volume of air is used, so that its temperature and degree of humidity can be determined.

The apparatus can be used for finding the velocity of sound in other gases than air by making the handle by which the piston is moved in the form of a narrow tube, and causing a slow stream of the gas to flow through it into the main tube. If a gas heavier than air is used, such as carbon dioxide, the tube should be placed with the open end C upwards; if hydrogen or coal gas is used, the end C should be downwards.

A doubt might arise as to whether the measurement of DB should be made from the nearer edge of A, the farther edge, or the centre. The correct place to measure from is the centre of the side tube. When the sound is at its minimum at A, what really happens is that the direct and reflected waves form stationary vibrations in CB, and A is at the centre of a vibrating segment, where there is motion but no change of pressure. Now it is only *exactly* at the centre of the segment that there is *no* change of pressure, and the tube A must take in a little on each side of the centre. If at any moment, however, there is compression on one side of the centre there is rarefaction on the other side, and so if the centre of the segment is opposite the centre of the side tube, the small regions on each side of the centre of the segment, which are also opposite the end of the tube A, will neutralize each other.

147. Zones of Silence round a Fog Siren. It has been noticed that when a fog-siren is placed on a high cliff, a ship may find itself in a "zone of silence," where the siren is inaudible; though if the ship move towards or away

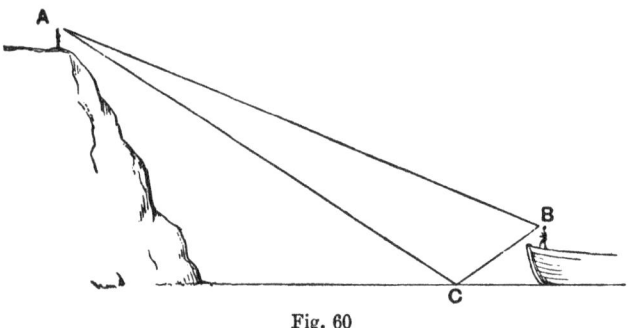

Fig. 60

from the shore the sound becomes audible again. These zones of silence have been explained by Tyndall as being due to interference.

Let A be the fog-siren, and B the observer on the deck of a ship. The sound can reach him by the direct line AB and also by the path ACB where it is reflected from the surface of the sea. If these paths differ by a half wave-length, or any odd multiple of a half wave-length, the two trains of waves will interfere at B, and the fog-siren will not be heard. If the ship moves either inwards or outwards, the difference between the two paths changes, and the sound is heard again. It is loudest when the difference of the paths is an exact number of wave-lengths.

148. Superposition of trains of waves of nearly the same Wave-length. We have so far considered only the interference of two trains of the same wave-length. There is a further case of great practical importance, where the wave-lengths, and consequently the frequencies, differ slightly.

Suppose we have two trains of waves travelling along the same line in the same direction, one of which has the frequency 100 and the other 101, and suppose that at some one moment they are in the same phase at a point A. At a point B whose distance from A is the distance sound travels in one second, they will also be in the same phase, for the length AB contains exactly 100 waves of the one train and 101 of the other. The waves with frequency 101 are a little shorter than those with frequency 100 and so, as we pass along the train from A towards B, the longer waves will gradually drop behind the shorter waves in phase.

If we call the wave-length of the shorter waves λ, then that of the longer waves will be $\lambda + \lambda/100$. Hence after passing over a distance λ, the train of longer waves will be $\lambda/100$ behind in phase*, after 2λ it will be $\lambda/50$, after 3λ it will be $3\lambda/100$, and so on. After passing over 50λ, that is half way from A to B, the difference in phase will be $\lambda/2$, or the two trains will more or less neutralize each other, according to

* The word Phase is not here used with its strictly correct meaning. This will not however cause any confusion.

their relative amplitudes. After passing the middle point C the longer waves will drop still farther behind, until at B they will be a whole wave behind, and so have come into the same phase again. We see then that at A the two trains will give a resultant wave with double the amplitude of either, if the two components have equal amplitudes, and the amplitude of the resultant will gradually fall off, as we move towards C. At C it will become zero and then will rise again, until at B it is the same as at A. As waves of all lengths travel with the same velocity, these maxima and minima will travel with the velocity of sound, and an ear placed in the path of the train will hear the sound rise and fall in intensity once a second.

We have supposed the two frequencies to differ by unity, but we can easily extend the proof, so as to include the case of two trains of waves whose frequencies have any difference. Let one train have frequency a and the other $a + n$. In the distance travelled by sound in one second there will be a waves of one train and $a + n$ of the other. The shorter waves gain on the longer by n waves. Every time one wave-length is gained, the trains come into the same phase, and there is maximum sound. Hence in passing from one end of the length v to the other, we pass over n maxima, with minima half way between them. It follows that as sound travels a distance v in one second, there will be a rise and fall in intensity n times per second.

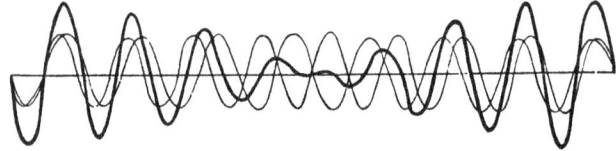

Fig. 61

Fig. 61 shews the change of amplitude of the resultant wave train, when the components have frequencies in the ratio of 8 to 9.

149. Beats. If the difference of frequency of the two notes is great, the alternations in intensity follow each other so rapidly that they cannot be distinguished, and we hear merely the two separate components sounding simultaneously. If the difference of frequency is small, or in other words the two notes differ little in pitch, we can no longer distinguish them separately. We hear merely a single note which rises and falls in intensity. These alternations in intensity arising from two notes of nearly the same pitch are called *Beats*, and play an important part in many acoustical determinations. By counting the beats in a given time we can find the number per second, and thus find the difference between the frequencies of the two notes.

The difference n need not be an integer. The general statement that the number of beats per second is the difference of the frequencies of the two notes holds for all values of n, whether integral or fractional.

150. Pitch of the note heard when two notes beat. It was said above that, when the notes are of nearly the same pitch, the ear hears only a single note with variations in loudness. The question arises *what* note does it hear? Is it the higher of the two components, or the lower, or something different from either?

The point is of no practical importance, and it is sufficient to say that the resultant curve is not harmonic. In so far as it can be said to have a wave-length, that wave-length is generally different at different points, and therefore the pitch fluctuates a little. At the maximum it is always intermediate between the pitches of the components. At the minimum the pitch depends on the relative intensities of the components. If they have nearly the same intensity, the pitch is intermediate, as at the maximum. If the lower note is much the stronger, the resultant has a lower pitch than either. If the higher note is much the stronger, the resultant has a higher pitch than either.

151. Illustrations of Beats. Beats are of very common occurrence and can be produced in many ways. Take two tuning-forks of the same pitch and flatten one of them by sticking a little wax on the end of each of the prongs. If they are now sounded together beats will be heard. Count the number of beats heard in as long a period as the forks continue to sound—say half a minute—divide by 30, and the quotient is the difference between the vibration numbers of the two forks.

Two organ pipes of the same pitch will beat strongly if the pitch of one is lowered a little by partly covering its mouth with the hand.

A pianoforte has generally three strings to each note, and the strings of any one note should be exactly in unison. If a number of notes be sounded in turn, it will generally be found that some of them give beats, shewing that the strings are not exactly in tune. The usual method of tuning a pianoforte is to get one of the three strings to the right pitch relatively to the other notes of the instrument, and then to bring the two other strings into unison with it by tightening them or loosening them until no beats are heard.

152. Limitation of the Method of Superposition of Waves. In all that has been said hitherto about interference and beats it has been assumed that we can get the resultant displacement at any point by adding the displacements that would arise from each of the two component trains separately. This is true only when the displacement is proportional to the force that causes it. We saw in Chapter II that the diminution in volume of air is not proportional to the increase of pressure, but if the variations of pressure are very small, as is generally the case, the deviation from proportionality is so small as to be negligible, and the conclusions we have reached may be regarded as being a correct statement of what takes place, if the sound is not very loud.

It might be anticipated however that, when the sound is loud, the defect of proportionality may be great enough to introduce new features, and this is found to be the case.

153. Combination Tones. Mathematical analysis shews that when two notes of frequencies p and q are sounded strongly together, a series of other tones called *Combination Tones* exist in addition to the two components.

The strongest of these tones is called the First Difference Tone, and has a frequency $p-q$. This tone may produce with the second primary a second difference tone, having the frequency $p-2q$. This again may produce a third difference tone of frequency $p-3q$, and so on. Also two difference tones such as $p-2q$ and $p-4q$ can theoretically produce a tone $2q$. We have also a First Summation Tone of frequency $p+q$, and other Summation Tones formed from this and each of the primaries, or from this and any of the Difference Tones. Thus an infinite number of Combination Tones are theoretically possible, and many of them can be derived in more than one way from the primaries and the other Combination Tones.

The First Difference Tone is generally the only one of the series that is easily audible. It is much weaker than the primaries, and therefore the Second Difference Tones, being formed from a strong tone and a weak one, are still weaker. The First Summation Tone is very weak. It can be heard faintly when two notes are sounded together very strongly on a harmonium, but there are observers who say they have never succeeded in hearing it, and who deny its existence. The First Difference Tone is easily heard on the harmonium. Sound together the note c^2 near the top of the treble clef and the note g^2 above it. The note c^1 will be heard as a Difference Tone. The tone will be heard more easily, if c^1 is sounded for a moment by pressing down its key, so as to prepare the ear for the note it is to hear. Difference Tones are produced in a peculiarly unpleasant form, when two tin whistles are blown loudly at the same time.

154. Theories of the origin of Combination Tones. Much discussion has taken place as to the origin of Combination Tones. The First Difference Tone has the same frequency as the beats of the two primaries, and some physicists have maintained that it is a subjective effect due to the beats. If this were the case we should be compelled to abandon Ohm's Law, for the superposition of two vibrations of different frequencies does not introduce a vibration whose frequency is the difference of the frequencies of the primaries, as we shall see more clearly when we discuss Fourier's Theorem. According to Ohm's Law a note is heard only when there is present in the air a harmonic vibration of the appropriate frequency, and a mere rise and fall of intensity is not such a vibration.

Other physicists have maintained that the combination

tones are formed in the ear or in the brain, and have no existence in the air. Rücker and Edser, however, have shewn that when two notes are produced simultaneously by a siren, their combination tones—both differential and summational —can excite resonant vibrations in a tuning-fork; whence we conclude that, in some cases at least, the tones have an objective existence.

The subject is too complex to discuss here at length. It must suffice to say that at present Helmholtz's Theory that combination tones are due to the failure of the Principle of Superposition is generally accepted as being correct in the main, though there are probably cases where Combination Tones arise from a want of symmetry in the mechanism of the ear.

155. Method of finding the pitch of Combination Tones. The difference tone arising from two notes forming any of the ordinary intervals used in music can be readily found from the table of harmonics in Fig. 31. Suppose, for instance, we want to find the first differential of two notes a major third apart. The eighth and tenth terms form the required interval. Their frequencies are in the ratio 8 to 10. If their actual frequencies were 8 and 10, the frequency of the difference tone would be 2. The fact that the frequencies are in reality about 65 times as great makes no difference, since intervals are measured by the ratios of frequencies. Hence the difference tone is number 2 of the series, that is, it is two octaves below the lower of the two notes which form the major third. This relation holds whatever is the actual pitch of the notes. So long as the primaries make an interval of a major third, the first difference tone is two octaves below the lower. We should have got the same result if we had taken numbers 4 and 5 to form the major third. The difference of their frequencies shews that number 1, the lowest member of the series, is their first difference tone, which is again two octaves below the lower primary.

To take another example let us find the first difference tone of $e^1\flat$ and $b^1\flat$, which are a fifth apart. There is no need to transpose the series of tones. The second and third members are a fifth apart. Their Difference Tone is number 1, which is an octave below the lower primary. Hence the first difference tone of $e^1\flat$ and $b^1\flat$ is also an octave below the lower of the two notes, or is $e\flat$.

The same method can be applied to find Summation Tones. The first Summation Tone of No. 4 and No. 5, which are a major third apart, is No. 9, which is a ninth above the lower primary.

It is a useful exercise to sound c^3 on the harmonium with each of the notes in the octave below it. As the lower note rises from c^2 through d^2, e^2 etc. the difference tone is heard falling from c^2 through $b^1\flat$, g^1 etc., and the tones heard can be compared with those obtained by the method of calculation just given.

It should be mentioned that the notes of a harmonium as ordinarily tuned are not quite in accordance with the ratios given by the harmonic series, and some of the difference tones will be found to be out of tune. The reason for this will appear when we speak of temperament.

156. Beats caused by Difference Tones. We shall have more to say about difference tones when we come to the subject of musical concords, but one instance of their use may be given here. A tuning-fork mounted on a resonance box gives a note which is practically a pure tone, that is, the vibrations produced are sensibly simple harmonic vibrations. Take two such forks an octave apart, and put one a little out of tune by sticking wax on its prongs. If the forks are now sounded together beats will be heard. The notes of the forks differ too much in pitch to beat directly, and we must look for another cause. Suppose one of the forks has a frequency 100 and the other 198. The first difference tone will have a frequency 98, and this gives two beats a second with the note of the lower fork. Two forks can be tuned to an exact octave by making use of these beats. If the forks are adjusted until the beats disappear, the ratio of the frequencies must be exactly 2 to 1.

CHAPTER VIII

RESONANCE AND FORCED VIBRATIONS

157. Free and Forced Vibrations. The vibrations of sounding bodies discussed in the preceding chapters are all of the kind known as free vibrations. They are the vibrations which a body executes if it is made to vibrate and then left to itself, and their period depends only on the dimensions and elastic constants of the body. The period of vibration of a body in such circumstances is called its *Free Period.*

We must now consider the case where a body is maintained in a state of vibration by a periodic force, which has not necessarily the same period as the free vibrations of the body.

When the period of the force is not the same as the free period of the body, the body ultimately vibrates in time with the force, and its vibrations are called *Forced Vibrations*. In the special case where the period of the force and the free period of the body are the same, we have the phenomenon known as *Resonance*. As the latter case is the simpler, we shall take it first.

158. Resonant Vibrations of a Pendulum. Make a simple pendulum by suspending a heavy bob by a string of such a length that the centre of the bob is 39 inches below the point of suspension. This pendulum will have a period of about two seconds, that is, at intervals of two seconds it will be found in a particular phase, say at the end of its swing to the right. Fix two light threads to the bob. Take one in each hand and, starting with the bob at rest, pull very gently to the right for one second, then to the left for one second, then to the right, and so on. By this means we apply to the pendulum bob a force which may be regarded as roughly representing a harmonic force, and it is clear that the bob will soon swing with considerable amplitude. The first pull will draw the bob a little to the right, the second pull will draw it to the left, the third will draw it rather farther to the right than before, the fourth still farther to the left, and so on. As the force changes its direction every second, and the pendulum changes the direction of its motion every second, the force will always act in the direction in which the bob would move if left to itself, and so the whole of the work done by the force will be used in increasing the amplitude of the swing, except such small amount as is required to compensate the loss of energy from air-friction, etc.

159. Forced Vibrations. If the pulls have a period that is not the same as the free period of the pendulum, the amplitude will not continually increase. Suppose, for instance, that the period of the force is a little less than that of the pendulum. For the first few swings the force will be nearly in time with the pendulum, and the amplitude will increase, but as the force gets more and more in advance of the pendulum in phase, a time will come when it is half a period in advance, and then it will act to the right when the pendulum is moving to the left, and vice versa, so that, instead of the force doing work on the pendulum and increasing its swing, the pendulum does work on whatever exerts the force, and for several swings the amplitude gets less.

These rises and falls of amplitude will continue for some time, but it can be shewn mathematically that the pendulum will ultimately settle down to a vibration that has the same period as the force. Such a vibration is called a *Forced Vibration*. We shall defer further consideration of forced vibrations to a later stage of this chapter, merely stating here that their amplitude is generally small, except when the period of the force is nearly the same as the free period of the body.

160. Vibrations produced by a Non-Harmonic Force. Suppose next that, instead of giving alternating pulls each lasting a second, we give every two seconds a pull of short duration and always in the same direction. The force is again periodic with a period

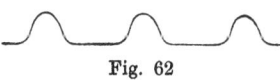

Fig. 62

of two seconds, but it bears little resemblance to a simple harmonic force. A simple harmonic force would be represented graphically by a sine curve, whereas the force we have just described would be represented by some such curve as that of Fig. 62. The result however will be the same as before. If the pendulum starts from rest it will gradually acquire a large arc of swing, for the force will always act at the right moment to increase the swing.

If the short pull is given every four seconds, we shall again have vibrations of large amplitude, and similarly if the interval between the pulls is any multiple of two seconds.

We see then that either a harmonic or a non-harmonic force can excite large vibrations in a body; but there is a difference between the two cases. A harmonic force excites resonant vibrations in a body only when its period is the same as the free period of the body, or one of its free periods if, as is generally the case, the body has more than one possible mode of vibration, each with its own period. The non-harmonic force *may* excite resonant vibrations when its period is not the same as the free period of the body. It is not correct to say that it *will* excite such vibrations, when its period is a multiple of that of the body, for it is only in special cases that it will do so. The discrimination between the cases in which a non-harmonic force of period different from that of the body does excite resonant vibrations and those in which it does not must be left until Fourier's Theorem has been explained.

161. Instances of Resonant Vibrations. Resonant vibrations are a common occurrence in everyday life. Every child knows that by certain motions of his body he can increase the arc of an ordinary swing, and he soon finds that his motions must keep time with the swing, if they are to be effective. If one leans first to one side and then to the other of a heavy boat, a considerable roll can be set up, if the motions of the body have the same period as the natural swing of the boat. A suspension bridge has been known to give way through the large swing produced by the regular tramp of soldiers, when the tramp happened to keep time with the natural swing of the bridge. The writer once experienced a similar resonant swing on a seaside pier. A large number of people were walking along it to the strains of a band, thus keeping in step with each other, and an oscillation was produced that was great enough to cause some persons to fall. Fortunately the alarm was so great that the people rushed about regardless of the band, and the oscillation quickly subsided.

Resonance can be shewn by making use of the monochord. Tune one of the strings to a tuning-fork. Make the fork vibrate, and hold its stem against one of the bridges. The string will quickly take up the vibration, as can be shewn by placing a paper rider on it.

It is not necessary for the fork to be in tune with the fundamental of the string. It will give resonance if it is in tune with any one of the overtones. If it is, for instance, an octave above the fundamental, the string will vibrate in two sections with a node between them. This is a convenient method of shewing the various modes of vibration of a stretched string.

162. Helmholtz's Resonators. The air in a hollow body with a narrow neck, such as a bottle, has a definite period of vibration, as is shewn by blowing across the neck, when a note of recognizable pitch is heard. If a tuning-fork of the same pitch as this note is held near the mouth of the bottle, the sound swells out greatly, through the resonant vibrations excited by the fork in the air.

Resonators similar in principle to this are of great use in acoustical investigations. They were used by Helmholtz in his work on the quality of musical notes, and are therefore usually called Helmholtz's Resonators.

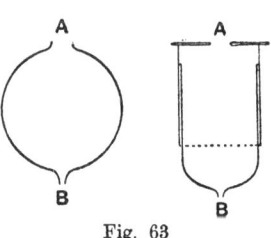

Fig. 63

Two of the more common forms of these resonators are shewn in Fig. 63. In each case A is the open mouth and B is a short neck which is inserted in the ear, so that, if resonant vibrations are excited in the contained air, they can be heard even though

132   RESONANCE AND FORCED VIBRATIONS   [CH. VIII

faint. The second form has the greater part of the body cylindrical and double, so that the volume of air can be adjusted by sliding the outer part over the inner.

These resonators are of such importance that we must consider their action in some detail.

163. Nature of the vibrations in a Helmholtz Resonator. We saw in Chapter II that the period of vibration of an elastic body is of the form

$$2\pi \sqrt{\frac{M}{F}},$$

where M is the inertia of the moving parts, and F is the elastic force. We cannot generally separate the body into two parts, and say that the inertia is all in one and the elasticity all in the other, but we can often see that the elasticity is mainly in one part, and the inertia mainly in the other. A bob hung by a light spring is an instance. The elasticity is wholly in the spring, and the inertia is mainly in the bob. There is some inertia in the spring, since it moves with the bob, but if it is much lighter than the bob its inertia is unimportant. We can make a similar approximate separation in the case of a resonator.

We may describe the motion of the air in a resonator in the following way. A stream of air runs in and out of the vessel by the neck A. When it runs in, it raises the pressure of the air inside. This rise of pressure checks the stream of air, and causes an outward stream. The momentum of the outward stream causes it to flow longer than is required to reduce the pressure inside to the same as that outside. The stream therefore overruns the equilibrium state, and the pressure inside the resonator sinks below that outside. Consequently an inward stream is produced again and the cycle of events is repeated.

Let us suppose the mouth A is small and the volume of the resonator is large; then when the air is streaming in and out, there will be little kinetic energy anywhere except near A. We may compare A to the centre of a loop in stationary vibrations and the whole inside of the resonator to a node. Near A there is a maximum motion and very little change of pressure, and inside the resonator there is change of pressure but very little motion. In other words the energy at the mouth is mainly kinetic, and that inside is mainly potential.

164. Dependence of the pitch of a Resonator on its Volume. The term F in the expression for the period of vibration of any elastic body is the force of restitution for unit displacement. In the case of the resonator it is the rise of pressure when unit volume of air is introduced. Now the rise of pressure resulting from the introduction of a given volume of air depends on the volume of the resonator, and if the volume introduced is small, the rise of pressure is inversely proportional to the volume S. Hence it follows that the period of the vibrations is directly proportional to $\sqrt{S}$.

We have now one method of tuning the resonator to a particular note. Decrease the volume to raise the pitch, increase it to lower the pitch. This is the reason for the slide in the second form of resonator in Fig. 63.

If the body of the resonator is so large and the mouth so small that there is no appreciable kinetic energy inside the resonator, the shape of the body is of no consequence. Take a tuning-fork and a bottle, and pour water into the bottle so as to alter the volume of the contained air, until it resounds most strongly to the fork. If the bottle is tilted, the shape of the part containing air is altered, but not its volume, and it will be found to give equally good resonance in all positions. This is not true when there is appreciable motion of the air inside the body. An organ pipe is a resonator, but it will be seen in a later chapter that a long narrow pipe and a short wide pipe of the same volume have not the same pitch.

The elasticity of all gases is the same except so far as they have different values for the ratio of their specific heats, so it does not matter what gas is inside the resonator so long as γ is unchanged, as the density of the gas is of no consequence where there is no motion. At the neck, however, the case is different, for here there is maximum motion, and the inertia of the moving gas is proportional to the density of the gas. We may conclude therefore that the period is proportional to the square root of the density of the gas near the mouth. Of course, the gas is generally the same at the mouth as inside, but it is of interest to see that, if the gas could be changed inside without changing that at the mouth, the pitch would be unchanged.

165. Dependence of the pitch of a Resonator on the size and shape of its Mouth. There is still one other feature which affects the pitch, and that is the size and shape of the mouth. The time taken for an excess of pressure inside to drive out the excess of air plainly depends on the size of the mouth. The more easily the air escapes, the shorter is the period of vibration. If we call the property of the mouth in virtue of which it allows the air to escape more or less easily its conductivity, we may infer that the greater the conductivity, the greater the frequency, or the higher the pitch.

The conductivity depends mainly on the size of the hole, but also to some extent on its shape. The calculation of its value is in most cases beyond the powers of mathematical analysis at present.

Rayleigh gives for the frequency of a Resonator

$$n = \frac{a}{2\pi} \sqrt{\frac{C}{S}},$$

where a is the velocity of sound in the gas, C is the conductivity of the opening and S the volume of the resonator. In this form of the expression for n the density and elasticity of the gas do not appear explicitly as they are both involved in the factor a.

166. Resonators with several Mouths. It is immaterial where the mouth is—there may in fact be several mouths without affecting the conclusions at which we have arrived. C will in this case be the sum of the conductivities of the separate channels.

If there are two equal openings so far apart as not to interfere with each other, the conductivity is double that of either opening separately, and therefore the frequency with two equal openings is $\sqrt{2}$ times that with one opening. This corresponds to an interval of nearly a fifth. If the openings are close together the interval will be rather less, for the conductivity depends not only on the size and shape of the hole, but also on the ease with which the stream of air can spread out when it leaves the hole.

The stream lines for a single hole will be somewhat as shewn in Fig. 64. For two holes close together they will be as in Fig. 65. The stream lines on the sides of the holes next to each other interfere with each other, and cannot spread as freely as in Fig. 64. Consequently the conductivity in Fig. 65 is less than twice that in Fig. 64, if all the holes

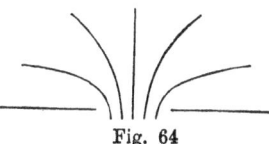

Fig. 64

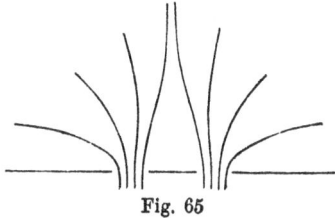

Fig. 65

are of the same size.

This argument can easily be extended to shew that a single hole of 1 sq. in. in area has a smaller conductivity than two holes each of $\frac{1}{2}$ sq. in. area, if the holes are all of the same shape; and further that a sheet of ordinary perforated zinc has greater conductivity than a sheet with a single hole, whose area is equal to the sum of the areas of all the holes in the perforated sheet*.

These conclusions as to the conductivity of openings have a practical bearing on methods of ventilation, for the ease with which the air can escape through a hole or a set of holes is a matter that has often to be considered. It is, for instance, sometimes useful to know that a long narrow opening has greater conductivity than a circular or square opening of the same area.

---

* This statement is not true if the holes in the perforated zinc are very small, for in that case the viscosity of the air becomes of importance and modifies the result.

167. Resonance Box of a Tuning-fork. When it is desired that a tuning-fork should emit a loud sound it is mounted on a resonator consisting of a wooden box with one end open, the dimensions being so chosen that the air contained in the box has the same frequency as the fork. The fork alone has such a small surface that it communicates very little energy to the air, and can only be heard when it is held near the ear, but when mounted on the box the contained air is set into resonant vibrations, and powerful waves are emitted from its mouth. It is evident that a fork mounted on such a resonance box cannot vibrate for so long a time as when it is merely held in the hand. The energy of the sound-waves comes from the energy of the fork, and when the waves are powerful, the drain on the fork is greater than when they are weak.

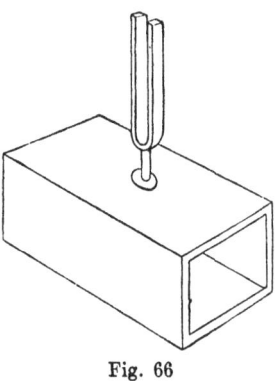

Fig. 66

The energy of air-waves is small, yet by virtue of the principle of Resonance the waves are able to excite powerful vibrations in elastic bodies that have the same natural period as themselves. This can be shewn with two such tuning-forks on resonance boxes as we have described. Call the forks A and B, and suppose they have the same pitch. Set A vibrating, and hold it near B for a short time. Now stop the vibrations of A, and B will be found to be vibrating. The fork A communicates its vibrations to its resonance box, this sends out waves which excite resonance in B's resonance box, the air vibrations in B's box make the wooden walls vibrate, and the vibrations of the walls start the fork B.

There is an old story that some celebrated singer with a powerful voice was able to break a wine glass by singing to it. If this really happened, it was no doubt a case of resonance. A wine glass has a natural period, as is easily shewn by tapping it, when a musical note is given out. If this note is sung, the glass will be set vibrating, and might conceivably vibrate to such an extent as to break.

168. Forced Vibrations. We return now to the case where the periodic force has not the same period as the free vibrations of the elastic body on which it acts. We have already stated that the amplitude of the vibrations excited will not in general be great.

The complete investigation of these forced vibrations is too complex to be given here, and only a few general results will be stated.

It can be shewn that if a body whose free vibrations have frequency p is acted on by a harmonic force of frequency n, the amplitude of the resulting vibrations of the body will be

$$A = \frac{f}{\sqrt{(p^2 - n^2)^2 + 4\kappa^2 n^2}},$$

where κ is the coefficient of damping of the body, or the constant which defines the rate at which the vibrations die away in consequence of viscosity, emission of sound-waves, etc. when the body is left to itself, and f is a constant which defines the intensity of the periodic force.

We see that the more the force and body differ from each other in period, the greater $p^2 - n^2$ will be, and the smaller the amplitude will be. In other words the more nearly the force is in tune with the free vibrations of the body, the greater will be the amplitude of the forced vibrations.

The maximum amplitude is obtained when $p = n$, for then $p^2 - n^2 = 0$. This is the case of Resonance, which we have already discussed at length.

It will be seen that if it were not for the damping there would be no limit to the amplitude of the resonant vibrations, for if $p = n$ and $\kappa = 0$, we have A infinite.

169. Effect of mistuning on the intensity of Resonance. It is not practicable to get the period of the force *exactly* equal to the free period of the body, and the question arises whether a given amount of mistuning has the same harmful effect on the amplitude of the resonance in all cases.

The mathematical investigation shews that the effect of mistuning depends on the damping, or the rate at which the free vibrations of the body die away. If they die away slowly, whether it is in consequence of the great mass of the body or the smallness of the frictional forces, the tuning must be close to get powerful resonance. The vibrations of a tuning-fork held in the fingers die away very slowly, and therefore, if it is attempted to set the fork in vibration by the action of airwaves, these waves must have very nearly the same frequency as the fork.

On the other hand the vibrations of the air in a resonator die away almost immediately the exciting force ceases to act. Consequently such a resonator gives a considerable amount of sound with a fork of a pitch anywhere near the proper pitch of the resonator. This can be tested with the second of the resonators shewn in Fig. 63. If the volume is gradually altered whilst a vibrating fork is held before the mouth, it will be found that there is resonance over a wide range of volume.

It is not difficult to see why the closeness of tuning necessary to give strong resonance must depend on the rate of damping. Resonance is due to the cumulative effect of the successive impulses. If the natural vibrations of the body die away quickly, there is little opportunity for the effect of the impulses to accumulate. Suppose, for instance, that the vibration of the body when left to itself would practically cease after 10 periods; then the effect of the first impulse might be regarded as having disappeared when the eleventh was given, and it would only be necessary to tune the force to the body so closely that they should remain approximately in the same phase for 10 periods in order to obtain almost the maximum amplitude possible. If on the other hand the body would perform 5000 vibrations before coming to rest, the tuning would have to be much closer to give strong resonance. If the tuning were only roughly approximate the force would come into a phase opposite to that of the body before the effect of the earlier impulses had disappeared, and so would neutralize the effect of those earlier impulses. In this case the period of the force should differ from that of the body by less than one part in 10,000 to get approximately the maximum resonance.

169–171] RESONANCE AND FORCED VIBRATIONS 139

*170. Phase of Forced Vibrations. The force and the resulting forced vibrations will not in general be in the same phase. The investigation of the difference between the phases is beyond our scope and nothing more than the results of mathematical analysis can be given here.

If δ is the difference of phase expressed as an angle, κ the coefficient of damping, p the frequency of the free vibrations of the body, and n the frequency of the force, then it can be shewn that

$$\tan \delta = \frac{2\kappa n}{p^2 - n^2}.$$

From this relation it is seen that the difference of phase between the force and the vibration varies with the amount of damping, and with the closeness of the tuning of the force to the body. If $p = n$ or the force is tuned exactly to the free vibrations of the body $\tan \delta$ is infinite. In this case, which is the case of resonance, the force and the vibrations differ in phase by a quarter of a period. If n is very small compared with p, $\tan \delta$ is very small, and the force and vibration nearly agree in phase. This is the case where the period of the force is much greater than the natural period of the body. In the opposite case where the body is constrained to vibrate much more rapidly than it would vibrate if left free, n is large compared with p, $\tan \delta$ is a small negative quantity, and the force and vibration therefore differ in phase by nearly half a period. We shall see in Chapter IX that Helmholtz made use of the dependence of the phase relationship on the closeness of tuning in his investigation of the cause of consonance.

*171. Initial stages of Forced Vibrations. We have spoken throughout of forced vibrations with the same period as the force being the *ultimate* result of the action of the force, but have said little of the initial stages. The most interesting case is that in which the period of the force is not greatly different from the free period of the body. In this case the force and the vibrations of the body do not differ much in phase during the first few vibrations, and the amplitude increases as in the case of resonant vibrations, the body vibrating in its own natural period. The phase of the force gradually gets more in front of, or behind that of the vibrations, and presently the phases become half a period different from each other. The force then checks the vibrations, which die down, to be increased once more when the phases come into coincidence again. These variations of amplitude continue for some time with a gradually diminishing range, and finally disappear, leaving only the forced vibrations of constant amplitude.

We can see from general considerations that something of the kind described will take place. During the earlier stages of the motion we may regard the body as performing simultaneously two vibrations of different periods. It has its own natural vibrations set up in the first few periods of the force, and it has also the forced vibrations, which have the same period as the force. These two sets of vibrations having different frequencies will give rises and falls of amplitude similar to beats. The free vibrations, however, are not maintained and gradually die away; whilst the forced vibrations have a constant amplitude determined by the greater or less degree of approximation of the period of the force to the natural period of the pendulum, and by the amount of damping. The variations of amplitude will thus disappear gradually, and nothing will be left except the forced vibrations.

172. Forced Vibrations used in Musical Instruments. Forced vibrations play an important part in musical instruments. We may take the pianoforte as an instance. The strings have so little surface that if they were merely attached to an open iron frame, very little sound would be heard. The air is not compressed before an advancing string and rarefied behind a retreating string, but to a great extent slips round from front to back of the string in time with he vibrations. The strings are therefore fixed on a sound-board to which their vibrations are communicated, and as the board has a large surface, it is able to pass on its vibrations to the air.

It should be observed that this is not a case of resonance, as the same sound-board has to serve for all the notes of the instrument. The sound-board has a small mass and a large

171-173] RESONANCE AND FORCED VIBRATIONS   141

coefficient of damping, and is therefore able to take up vibrations over a wide range of pitch. If the coefficient of damping were small, any notes of the pianoforte that lay near the natural tones of the sound-board in pitch would be unduly loud. The natural tones of the sound-board can be distinguished immediately after a note is struck, but they die out after 4 or 5 vibrations, and leave only the forced vibrations.

173. Action of the Sound-Board. The manner in which a string communicates its vibrations to the sound-board sometimes presents a difficulty to the student.

Let AB (Fig. 67) be a vibrating string attached by two

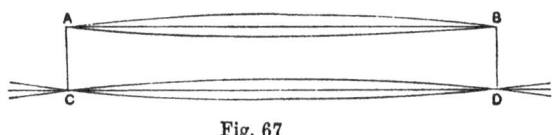

Fig. 67

pins to the sound-board CD, and suppose the string is vibrating in the plane of the paper. The effect of the motion of the string is to draw the pins a little towards each other, when it is at the upper and at the lower limit of its swing. The pins therefore vibrate with half the period of the string, and it might be thought that the sound-board would vibrate with the period of the pins, and give out a note an octave higher than that of the string. The sound-board however follows the motion of the string. When the string is at its upper limit, the pins are drawn together, and the sound-board is bent into a curve which is concave upwards. When the string moves downwards to its equilibrium position, the sound-board straightens again, and then in consequence of its momentum takes the shape of a curve concave downwards. Meanwhile the string is moving below its equilibrium position and again drawing the pins together. This increases the bending upwards of the sound-board, or increases the amplitude of its swing. The process is continued, and the period of the vibrations of the sound-board is thus the same as the period of the vibrations of the string, since sound-board and string are always moving in opposite directions. If by some

142   RESONANCE AND FORCED VIBRATIONS [CH. VIII

means the pins were made to vibrate in a direction at right angles to the plane of the board, the pins and board would have the same period.

The difference between the two cases can be illustrated by Melde's experiment.

174. Melde's Experiment. A light string has one end fixed to the prong of a tuning-fork as shewn in Fig. 68. The other end passes over a pulley and carries a scale-pan.

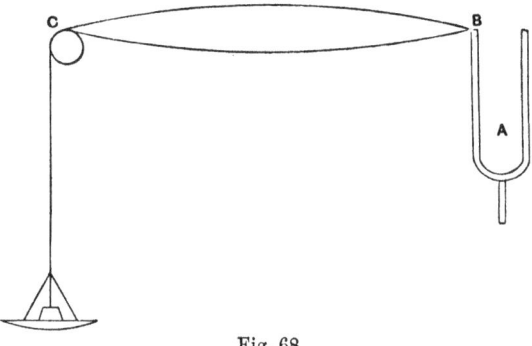

Fig. 68

Set the fork vibrating, and put weights in the scale-pan until the string vibrates in a single segment as shewn in the figure.

Now turn the fork round until its prongs vibrate in a plane at right angles to the string, and the string will be found to vibrate in two segments.

In the second case the end B of the string is drawn from side to side by the fork, and it is clear that the period of vibration of the string must be the same as that of the fork. It follows that in the first case, where the vibrating segment is double the length of that in the second, the string must vibrate with a period which is twice the period of the fork. The explanation is the same as was given for the action of the pins on the sound-board. Whilst the prong of the fork performs one complete vibration from its extreme left-hand position to its extreme right, and back again to its extreme left, the string moves only from its extreme position in one direction to its extreme position in the other direction. The string executes one complete vibration whilst the fork executes two, or the period of vibration of the string is twice that of the fork.

175. The Ear. Both forced and resonant vibrations probably play a part in the mechanism by which the ear carries sound from the external air to the brain, but the action of the ear is not fully understood. The short external passage is closed at its inner end, by a stretched membrane called the *tympanum*. Behind the tympanum is a small air-space called the *middle ear*, connected with the back of the mouth by the Eustachian tube, and so maintained at atmospheric pressure. At the inner end of this air-space is the *cochlea*, a narrow tube closed at both ends and coiled on itself like a snail-shell. A chain of three small bones across the middle ear connects the tympanum with a membranous window at one end of the tube of the cochlea.

Air-waves entering the external passage cause forced vibrations in the tympanum. These vibrations are conducted by the chain of bones to the window in the cochlea, and thence to the fluid contents of the cochlea. The cochlea is divided in two for almost the whole of its length by a partition. This partition is bony for the greater part of its width, but a narrow strip adjoining the wall of the cochlea is membranous and is called the *basilar membrane*.

It is believed that the basilar membrane enables us by its resonance to distinguish between sounds of different pitches.

The membrane increases gradually in width as we pass inwards round the coils of the cochlea. It is tightly stretched across its width but unstretched longitudinally. It may thus be compared to a large number of stretched strings placed side by side, the strings varying in length, and perhaps also in tension. Each narrow strip across the width of the membrane will have a definite period of vibration, and will resound only to such vibrations in the fluid of the cochlea as have that period. If then several notes of different pitch enter the ear simultaneously, each will excite its own particular part of the membrane, and thus by means of nerves running from the various parts of the membrane to the brain we are able to distinguish between the sounds. If the theory that has been given is correct—it is by no means certain that it is so—the basilar membrane serves as what might be called the acoustical equivalent of a spectroscope, in that it resolves complex air-vibrations into their simple harmonic constituents.

CHAPTER IX

QUALITY OF MUSICAL NOTES

176. Meaning of the term Quality. We mean by *Quality* the characteristic which enables us to distinguish a note sounded on one instrument from a note of the same pitch sounded on another instrument. The instruments used in an orchestra are chosen so as to give variety of quality, and much of the charm of orchestral music arises from the contrasts of sound-quality.

It is no doubt true that, when listening to an orchestra, the notes of the instruments are recognized to some extent by other characteristics than their quality. The percussion instruments such as the drum, harp, or pianoforte give notes which die away quickly, whilst the notes of the wind and stringed instruments can be maintained unchanged in intensity. There are slight differences too in the beginning or "attack" of the notes. Some instruments begin their notes explosively, others reach the desired intensity by a rapid *crescendo*. The main difference however is in quality. There is no difficulty in distinguishing the note of a hautboy, for instance, from that of a flute, or a violin, though the pitch and intensity are the same.

In Chapter I we were led to surmise that differences of quality are related to differences in the "shape" of the sound-waves. We shall now examine this surmise in detail.

177. Resolving power of the Ear. When two or more notes of different pitch are sounding together, the vibrations in the air due to each of the notes separately are compounded into a complex vibration. The ear is in general able to analyse this complex vibration, and to distinguish the separate notes. There is a marked difference between the ear and the eye as regards the power of analysing complex vibrations.

Light consists of waves in the hypothetical ether, and trains of ether waves of different wave-lengths correspond to lights of different colour. If lights of different colours strike the eye together, they are not distinguished separately; the eye sees only some one colour. The light from a Geissler tube containing hydrogen, for instance, is known from spectroscopic observations to consist mainly of red, green and blue rays, yet the unaided eye sees only a reddish light. Also the same effect may be produced on the eye by different mixtures of colours.

The eye needs a spectroscope to enable it to decide what are the constituents of a beam of light. The ear contains within itself the acoustical equivalent of a spectroscope. It is able with a little training to distinguish in most cases the constituents of a mixture of sounds of different pitch, though, as we shall see presently, the analysis may be made easier and more certain by the use of instrumental appliances.

178. Limitations to the resolving power of the Ear. The above statement as to the analysing power of the ear does not hold when the constituent notes are very close to each other in pitch. In such a case we hear a single beating note when there are only two constituents, and a confused noise without definite pitch when there are many. We may regard this, not as a failure of the general statement, but as a case where we are approaching so near to the limit, where the constituents have the same pitch, that the mechanism of the ear is not delicate enough to distinguish between them. We have the same apparent failure in optical instruments. The light of the glowing vapour of sodium consists mainly of two kinds with nearly the same wave-length and colour. A small spectroscope shews only a single line in the spectrum—its resolving power is not sufficient to separate the two lines. If we wish to see the two separately, we must use an instrument of higher resolving power. Similarly the ear has a limited resolving power, in that it ceases to be able to appreciate separately two notes sounding together, when the interval between the notes is small.

There is a further exception to be made. A note must have a certain minimum intensity before it can be heard at all in the presence of other notes. A note is easily drowned by another of lower pitch, but a note of low pitch is not so easily drowned by one of high pitch.

Admitting then the general statement that the ear can analyse the complex vibrations resulting from the simultaneous production of several notes of different pitches, the question arises, what kind of vibration in the air gives the sensation of a pure tone, recognized by the ear as having only a single definite pitch, and incapable of analysis?

179. Ohm's Law. The answer to this question is given in Ohm's Law, which states that a Simple Harmonic Vibration is the only kind of vibration that gives to the ear the sensation of a pure tone.

The law rests to a great extent on our belief that the mechanism by which the ear analyses a note is similar to that which we find in dead matter. A complex note can be analysed by refraction or other processes, and in every case the analysis is in accordance with Ohm's Law. The constituents separated out are always simple harmonic vibrations.

The vibrations of a stretched string provide an experimental proof of Ohm's Law. If a string is plucked at a particular point, it is possible to find mathematically the nature of the vibrations produced, and to express these vibrations as the sum of a number of simple harmonic vibrations, whose periods can be calculated.

It is found that the tones recognized by the ear correspond in pitch with the calculated periods. If calculation shews that a harmonic constituent of a particular frequency is present in the complex vibration, the corresponding note is heard; if the vibration is absent, the note is not heard.

180. Displacement Curves of Complex Vibrations. Musical notes are seldom, if ever, quite pure. The note of a tuning-fork mounted on a resonance box is almost a pure tone, as is also the note of a closed organ pipe of wide bore, but the great majority of musical notes are complex, and can be analysed into a number of constituents.

Let us suppose that a note contains only two tones an octave apart. We can find by the method described in

Chapter IV the displacement curve for the complex vibration, if we know the amplitudes and relative phases of the two constituents.

Suppose the amplitudes and phases are as shewn in the two upper curves of Fig. 69. Then by adding the ordinates

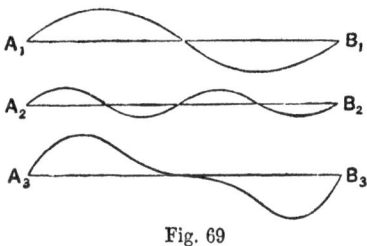

Fig. 69

and taking account of their signs in the usual way we get the lowest curve for the displacements of the air when the two notes are sounded together.

It will be noticed that the curve is not symmetrical on the two sides of the maximum ordinates. If, as usual, ordinates measured upwards correspond to displacements of the air particle forwards, or in the direction in which the wave travels, we see that any particle moves forward quickly and returns more slowly.

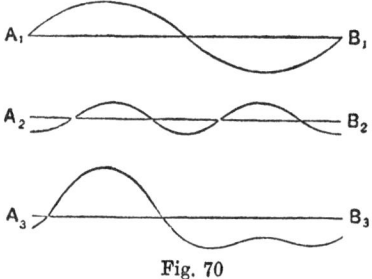

Fig. 70

Next take two constituents of the same wave-length and amplitude as before but with a different relation to each other as regards phase. Move the second curve of Fig. 69 to the right by a quarter of its own wave-length, and we get Fig. 70. The compounded curve is quite different from that of Fig. 69. The curve is now symmetrical on the two sides of its maximum ordinate.

In Figs. 69 and 70, we have drawn only one complete wave in the upper curve. This curve may be imagined to be continued indefinitely to the right and left, so as to give a train of waves, and similarly the second curve in each set can be continued in each direction. Whatever the relation of their phases may be, one wave of the upper curve is compounded with exactly two waves of the second curve, and therefore the two curves have the same relation of phase at B as they have at A. The train of waves of which A_1B_1 is a part will consist of a series of repetitions of A_1B_1, the train of the second line will consist of repetitions of A_2B_2, and therefore the compounded train will consist of a series of repetitions of A_3B_3.

181. Periodic Curves. A curve which repeats itself in this way is called a *Periodic Curve*, and the length of the projection on the axis of the shortest piece of the curve that is repeated is the wave-length. The wave-lengths of the first and third curves are A_1B_1 and A_3B_3 but the wave-length of the second is $\frac{1}{2}A_2B_2$, since the right-hand half of A_2B_2 is a repetition of the left-hand half. Call the wave-length of the upper curve λ, then we may state our result thus. If a simple harmonic curve, or a sine curve of wave-length $\frac{1}{2}\lambda$, is superposed on a sine curve of wave-length λ, the resultant curve is a periodic curve of wave-length λ, and this resultant curve has different shapes with different relations between the amplitudes and phases of the two constituents.

Suppose next we superpose on the curve of wave-length λ one of wave-length $\frac{\lambda}{3}$. A_2B_2 will now contain three waves to one in A_1B_1. We shall again have a variety of curves whose shapes depend on the relations between the amplitudes and phases of the two constituents, but which will all be periodic with a wave-length λ.

Further, the $\frac{\lambda}{2}$ and $\frac{\lambda}{3}$ curves could be superposed on the λ curve at the same time, and we should still have a resultant periodic curve of wave-length λ.

This process can evidently be carried on indefinitely by superposing curves of wave-lengths $\frac{\lambda}{2}, \frac{\lambda}{3}, \frac{\lambda}{4}$, etc., with any relative amplitudes and any relations of phase, and we shall always get as resultant a periodic curve of wave-length λ.

Thus, it appears that we can get an infinite number of trains of waves of different shapes but all of the same wave-length by compounding curves with wave-lengths λ and its aliquot parts. Is there any limit to the shapes of curves we can get in this way? Suppose we draw any arbitrary curve,

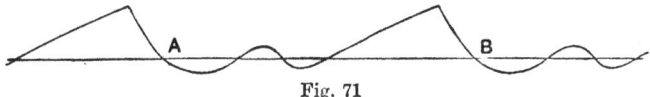

Fig. 71

such as that in Fig. 71, where the wave-length is AB. Can we build up this curve by compounding sine curves of wave-length AB and its aliquot parts?

182. Fourier's Theorem. The important theorem known as Fourier's Theorem states that *any periodic curve whatever* of wave-length λ can be built up by compounding sine curves of wave-lengths $\lambda, \frac{\lambda}{2}, \frac{\lambda}{3}, \frac{\lambda}{4}$, etc., with two limitations:—

(1) The curve must not overhang anywhere as at A in Fig. 72, and

(2) It must not have any infinite ordinate as at B.

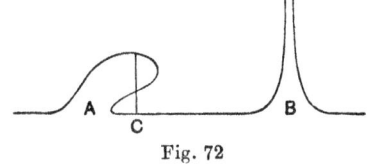

Fig. 72

If we are dealing only with sound-waves, these limitations

do not concern us, for each represents an impossible condition in the air or other medium. An overhanging piece of the curve as at A would mean that at some such point as C there would be two ordinates of different lengths, and this would require that a given particle of air could have two different displacements at the same moment. Limitation (2) clearly has no application to sound-waves, for a particle of air cannot have an infinite displacement.

Fourier not only shewed that any periodic curve could be built up from sine curves, but he also devised a method by which the constituents of any periodic curve could be found. The proof of the theorem and the method of employing it to analyse a curve require considerable mathematical attainments, and nothing more than general results can be given here.

The theorem can be stated in various ways. Up to this point we have treated it as a mere geometrical proposition applied to curves, but we have seen that curves can be used to give a graphic representation of vibrations, and we can therefore transfer the theorem from the curves to the vibrations represented by the curves.

If we take our curves to represent the displacements of a particle of air, when a train of waves is passing over it, we may say—since the period of vibration of the particle is the time taken by the train to travel one wave-length, and therefore the period is proportional to the wave-length—that any periodic vibration with a period τ can be represented as made up of simple harmonic vibrations of periods τ, $\dfrac{\tau}{2}$, $\dfrac{\tau}{3}$, etc.

These vibrations are what we found for the possible modes of vibration of a stretched string, when we have 0, 1, 2, 3, etc. nodes between those at the ends, and they correspond to the tones of the harmonic scale given in Fig. 31, so that this scale now acquires additional importance, for it enables us to state Fourier's Theorem in yet another way, which is the most useful for our present purpose.

Any complex musical note, whose frequency is n, may be treated as made up of a series of pure tones of the harmonic scale with frequencies n, $2n$, $3n$, etc.

183. Applications of Fourier's Theorem.

The terms needed to build up any given curve may be limited, or may be infinite in number. Any term, including the first term of the harmonic series, may be missing. Take for instance the case of a complex vibration due to two tones near enough in pitch to give beats. If the tones are pure they arise from harmonic vibrations, and Fourier's Theorem cannot extract from the compound vibration anything more than is put into it. Suppose the two tones have frequencies 100 and 101, then Fourier's analysis gives these two vibrations and nothing more. How are these to be accounted for as members of the Harmonic Series? The answer is that they are the 100th and 101st terms of the series, and no other terms are present. The frequency of the compounded vibration is 1, for in one second, and not earlier, each of the constituents will have completed an exact number of vibrations. The fundamental then in this case has unit frequency but zero amplitude.

If we compound two harmonic vibrations whose frequencies are incommensurable, such as 100 and 100π, or $100\sqrt{2}$, Fourier's Theorem does not apply. The compounded vibration is not periodic, as the initial circumstances will never recur, and the theorem applies only to periodic vibrations or curves.

When the complex curve to be analysed has only slight curvature everywhere, the lower members of its constituent harmonics have relatively the greatest amplitude. When there are points on the curve where the curvature is great, the higher terms become more important, and when the curve has sharp corners, such as those in Fig. 71, it can only be expressed by an infinite number of terms of the series.

Stating the same thing in terms of vibrations, we may say that the more sudden are the changes of velocity of a particle in the medium, the stronger are the higher constituent harmonics as compared with the lower. This feature has applications to the construction and use of musical instruments, and will often be referred to in the following chapters.

We need not stop to inquire whether it is possible mathematically to build up a complex periodic curve from some other fundamental curve than the sine curve, for Ohm's Law tells us that the ear analyses notes into simple harmonic constituents, and therefore our mathematical analysis must follow the same method.

The sine function is by far the simplest periodic function with which we are acquainted. In many different branches of physics Nature's method of analysis of periodic changes is in accordance with Fourier's Theorem. Whenever a phenomenon can be represented by a periodic curve, the theorem can be used to analyse the curve, and find what harmonic constituents it possesses. The theorem is in constant use for analysing such phenomena as the prevalence of sunspots and magnetic storms, the rise and fall of potential in electrical machines, the rise and fall of tides, etc.

184. Nomenclature. There is a want of uniformity in the terms used by different writers to designate a complex note and its constituents. We have used the following nomenclature up to this point, and shall continue to use it.

The complex sound made up of two or more constituents of different pitches is termed a *note*; and each of the constituents is called a *tone*. A tone is therefore the simplest constituent of a sound and is incapable of analysis.

When a body is capable of vibrating freely in several different modes, as we have seen to be the case with a stretched string, the lowest tone that it is capable of giving is called its *fundamental*, and the others are called *overtones*. The various modes of free vibrations of a body are often also called the *proper modes* or *natural modes* of vibration of the body.

The overtones of musical instruments in many cases have pitches in accordance with the harmonic series, Fig. 31. In such cases the overtones are called *Harmonic Overtones*.

The student must be careful not to confuse Overtones with Harmonic Overtones. The overtones of organ pipes diverge more or less from the Harmonic Series according to the width of the pipe, and the overtones of drums, tuning forks, and some other instruments bear no relation to the Harmonic Series. When it is desired to specify such overtones, they are called *Inharmonic Overtones*.

The constituents into which a periodic vibration is analysed by Fourier's method are generally termed *Harmonics*. Some writers use the term *Partials*, or *Harmonic Partials* for the Fourier constituents.

185. Helmholtz's Theory of Quality. Helmholtz was the first to make a full investigation of the causes of the differences between the qualities of the notes given out by different instruments, and he stated his conclusion as follows:

The quality of a musical note depends only on the number, order, and relative strengths of its harmonic constituents and not on their differences of phase.

This is in accordance with our surmise in Chapter I that quality depends on wave-form, but with the limitation that differences of form which are caused only by change of the phase relations of the constituents do not give differences of quality. According to Helmholtz's view, the third curves in Figs. 69 and 70, though different in shape, would correspond to the same quality, as they differ only in the relative phases of the two constituents.

In order to prove his theory Helmholtz first analysed experimentally the notes of several instruments to find the intensity of the various harmonic constituents, and then by means of tuning-forks he sounded these constituents together with the proper intensities, and found that he reproduced the original quality.

186. Analysis of Complex Vibrations. It is possible with a little experience to analyse a note into its harmonics to some extent by the unaided ear, but such analysis is rendered difficult from our habit of disregarding the harmonics of a note, and concentrating our attention on the fundamental.

The note of a violin, for instance, consists of the same harmonics in much the same proportions whatever its pitch, and we have accustomed ourselves to fuse together the separate sensations due to the various harmonics into a single sensation, which we call the quality of the note of a violin; separating out only that harmonic which has the lowest pitch, and calling it "the pitch" of the complex note.

Some of the harmonics of the note of a pianoforte can be heard without much difficulty. Strike the note C_1 loudly, and hold down the key, so as to keep the damper off the strings. The lowest 7 or 8 harmonics can generally be heard, if the attention is directed to each in turn. The detection of a given harmonic is made easier, if the corresponding key is touched gently before the note C_1 is sounded, so as to prepare the ear for the note to which its attention is to be directed. The fifth harmonic e is often heard to be surprisingly strong, when the ear has once isolated it. A person with a trained musical ear will notice that this fifth harmonic of C_1 is a little flatter than the e of the pianoforte. The notes of a pianoforte are tuned to equal temperament, which makes all the major thirds a little sharper than the just intonation, as will be described later. The difference of pitch between the two notes is easily detected, if C_1 is first sounded strongly, and then whilst its key is held down, the e of the pianoforte is sounded gently. Beats will then be heard from the want of coincidence of pitch between the fifth harmonic of C_1 and the e of the pianoforte.

The detection of the harmonics is made easier if resonance is used. Depress gently the key corresponding to one of the harmonics of C—say g—so as to raise its damper but not to sound the note. Now sound C loudly and hold down its key. The third harmonic of C will cause resonant vibrations in the strings of g, and if after a few seconds the finger is taken off the key C, the note g will be heard.

Helmholtz analysed the notes of musical instruments with the help of the resonators shewn in Fig. 63. He used a set of resonators tuned to the harmonics of B♭. Whilst a trombone, for instance, was blown to give the note B♭, he held the resonators to his ear in turn and recorded his estimate of the relative strengths of the various harmonics. In the same way he found what harmonics are present, and with what intensities, in the notes of other instruments.

The next step was to try to imitate the quality of the note of any particular instrument, by producing pure tones corresponding in pitch and intensity to those found by analysis in the note of the instrument, and sounding them together.

187. Electrically maintained Tuning-forks.

The most convenient way of producing a tone approximately

pure is by holding a tuning-fork of the proper pitch before a resonator, and this was the method employed by Helmholtz,

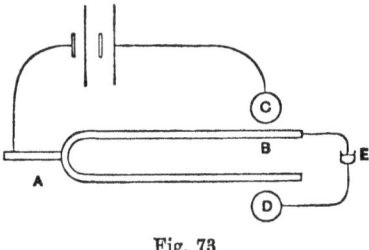

Fig. 73

but as the vibrations of a tuning-fork soon die away, an electrical method was used for maintaining them.

A tuning-fork has attached to the prong B a bent wire which just touches the surface of the mercury in the vessel E, when the fork is not vibrating. The ends of the prongs of the fork are between the poles C and D of an electromagnet. When the wire makes contact with the mercury at E, an electric current flows from the battery through the coils of the electromagnet, from this to the mercury in E, and so by way of the fork back to the battery. The electromagnet is thus excited by the current, and the prongs of the fork are attracted outwards towards the poles, but as soon as the upper prong has moved a little way the contact breaks at E, the circuit is broken, and the magnet loses its magnetism, allowing the prongs to fall back. The circuit is then closed again and the prongs of the fork drawn apart as before, and this process is repeated continually, the fork being thus maintained in a state of vibration.

Whilst the vibration is going on, the spark produced when the current is broken makes a good deal of noise, and so this fork cannot be used for our present purpose. It will be seen that the current in the circuit is interrupted once during each vibration of the fork, and this periodic current can be made to drive other forks placed so far from the interruptor as to be out of hearing of the crackle of the spark.

QUALITY OF MUSICAL NOTES

The fork whose vibrations are to be maintained is placed with its prongs between the poles of an electromagnet, and the intermittent current produced by the interruptor is passed through the coils. The fork will now have a periodic force acting on it, and will be made to execute forced vibrations with the period of the force. If the natural period of the fork is the same as that of the current, or in other words the interruptor fork and the maintained fork have the same pitch, there will be one impulse for each vibration, and by the principle of resonance the vibrations will soon become large.

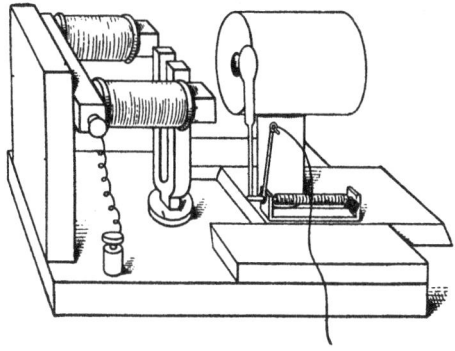

Fig. 74

Similarly, if the maintained fork has twice the frequency of the interruptor fork there will be one impulse for each two vibrations, and in this case also large resonant vibrations will be produced, and so on; any fork whose frequency is an exact multiple of that of the interruptor having large vibrations produced in it.

The action of the intermittent current on the forks may be explained in another way. When the interruptor fork has reached a steady state, the current must be periodic, and therefore can be regarded as made up of a series of currents which vary harmonically in intensity, and have frequencies 1, 2, 3, etc. times that of the interruptor. If the maintained forks are tuned to these frequencies 1, 2, 3, etc., each will be in tune with one of the harmonic constituents of the force acting on it, and so will be maintained in active vibration. This is a case of the general principle that a periodic force will produce large vibrations in a body on which it acts, provided one of the harmonic constituents of the force has the same period as one of the natural vibrations of the body. This principle should be specially noted as it has important applications to the theory of musical instruments.

The make and break of the current is fairly sudden; consequently the curve by which it could be represented has points of great curvature at the make and break. From what was said before about the Fourier analysis of curves with great curvature, it follows that a long range of terms of the harmonic series will be required to express the periodic force, and therefore forks with pitches high in the harmonic series can be maintained by it.

If the make and break were quite sudden, and always occurred at the same point in the swing of the prongs, the interruptor fork would not be maintained in vibration; for the prong would have the same force acting on it in its outward and inward journeys, and over the same range in each, and so would lose as much energy in one half of its vibration as it gained in the other. Both make and break however are delayed a little by two causes. The self-induction of the circuit, which is considerable on account of the presence of the electromagnet, causes a spark at the break, which prolongs the current a little; it also prevents the immediate rise of the current to its full strength at the make. Further, the wire sticks to the mercury at the break, and draws it up a little before the wire and surface part company, and at the make the surface is not broken and contact made until a dimple has been formed in the surface. The result is that the work done by the magnetic force while it acts in the direction of motion of the prong is a little greater than that done while it acts against the motion, and so, on the whole, energy is communicated to the fork.

188. Helmholtz's Synthesis of Complex Vibrations. Helmholtz used 8 forks tuned to the first 8 terms of the harmonics of B♭. Behind each fork was a resonator, which could be placed at different distances from the fork, so as to intensify the sound by any desired amount, and each resonator had a shutter by which its mouth could be closed (Fig. 74). The electromagnets of the 8 forks were all placed in series in the electric circuit of the interruptor fork and so all the forks were kept in vibration.

This method of driving the forks not only permits the noisy spark to be put where it will not be heard, but has the further advantage that it ensures the frequencies of the notes given out being exactly in the ratios 1, 2, 3, etc. If one of the forks is a little out of tune when vibrating freely, the current forces it into the proper pitch, and the only harm that is done is that the amplitude of its vibrations is not quite so great as it would be if the fork were exactly in tune. Since a tuning-fork without a resonator is almost inaudible, very little sound is heard when the shutters of all the resonators are closed.

Helmholtz found by analysis with his resonators that the note of a particular organ pipe consisted mainly of three members of the harmonic series, No. 1 strong, No. 3 moderate, and No. 5 weak; the other members being inaudible. He then tried to build up this note with his forks. The shutters of all the resonators were closed except those of Nos. 1, 3, and 5, and these three resonators were separately adjusted to such distances from their forks as to give tones of the same relative intensities as had been found in the analysis of the note of the organ pipe. On allowing the three forks to sound together with this adjustment of intensities, it was found that the quality of the note of the organ pipe was reproduced very closely. In the same way the notes of the horn, clarinet, and some other instruments were imitated. The notes of the hautboy and violin could not be reproduced with so few forks as 8. The penetrating quality of these instruments arises from the prominence of high harmonics in their notes, and many of these harmonics lay beyond the range of Helmholtz's forks.

189. Relation between the quality of a note and the phases of its constituents. Helmholtz used this set of forks also to investigate the connexion between the relative phases of the constituent harmonics and the resulting quality of note produced.

It was stated in Chapter VIII that, if a resonator be put slightly out of tune with a fork, the resonance will be weakened, and at the same time the phase of the resonator vibrations will be altered.

We have seen that the pitch of a resonator can be lowered by making the mouth smaller, and this was the method used by Helmholtz. He partly closed the mouth of a resonator, thus putting it out of tune with its fork. This altered the phase of the vibrations, and at the same time diminished the resonance. He then moved the resonator nearer to the fork, and so restored the intensity of the sound. The harmonic corresponding to this fork had then the same pitch and intensity as before, but a different phase in relation to the phase of the current. He found that no difference was made in the quality of the compound note by altering one or more of the constituents in this way, and therefore concluded that the phase of the harmonics has no effect on the quality.

This conclusion is not accepted by all physicists, and much has been written on the subject since Helmholtz's time. It would perhaps be correct to say that at the present time the prevalent opinion is that Helmholtz's theory is right as a first approximation, but that change of relative phase of the harmonics of a note is not quite without effect on its quality.

190. Speech. Differences in the quality of sounds play an important part in ordinary speech. Consonants are in many cases merely methods of beginning and ending vowel sounds. They are only passing sounds and not continuous. The vowels on the other hand are musical sounds which can be maintained indefinitely.

Take the word *bad* as an instance. If in the course of a song this word has to be sung on a particular note, and the note has to be maintained, say over a semibreve, what happens is as follows. At the beginning of the semibreve the voice begins with a special kind of explosion which represents the *b*. It then settles down to the vowel *a*, and maintains it to the end of the semibreve, when the note comes to an end with another kind of explosion, which represents the *d*. Neither the sound of *b* nor that of *d* can be maintained. It is only a vowel that can be maintained in singing. Consonants are in most cases mere noises, whilst vowels are musical notes.

Moreover, a vowel maintains its characteristics so long as it is maintained. It is not necessary to hear the beginning or end to decide what vowel it is. It is clear then that our recognition of any vowel must be due to its quality. We distinguish between the vowel $\bar{a}$ and the vowel $\bar{u}$ in the same way as we distinguish between the sound of a violin and that of a flute.

191. The Vocal Organs. The voice is produced by forcing air from the lungs through the opening between a pair of stretched membranes, each of which is able to cover half of the larynx or passage from the lungs to the mouth. These membranes are called the Vocal Chords, and when not in use for speaking or singing, their free edges are widely separated, so as not to interfere with the breathing.

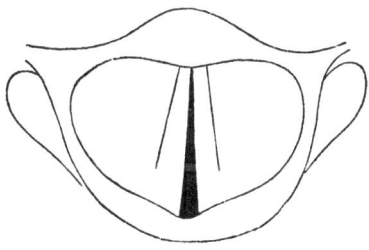

Fig. 75

By means of muscles these membranes can be stretched and their edges brought together. If air is now forced between them, their edges are set in vibration, and the air issues in a series of puffs, which give rise to a musical note.

The pitch of the note is varied mainly by altering the tension of the chords, the changes of tension being brought about by muscles attached to the larynx. The pitch and quality of the note can probably also be altered by changes in the distribution of the mass of the chords. On their under surface there is a layer of membrane, which can be moved towards or from the edge, thus weighting the vibrating part to a greater or less extent, and so altering the period and nature of the vibrations. The adjacent edges of the chords probably touch each other in the course of each vibration, and so make the stream of air discontinuous. The result of this discontinuity is that Fourier's analysis gives a long series of harmonics in the note produced. It is possible to detect as many as 15 or 16 in the note sung by a powerful bass voice.

The sound has to pass through the mouth on its way to the outer air, and the mouth and its adjoining cavities have natural periods of their own. Consequently such harmonics as approximate in pitch to any of the natural periods of the mouth will be strengthened by resonance, and the quality of the note will be altered. The pitch of the mouth regarded as a resonator can be altered at will, either by altering its volume, or by altering the size of its opening. Changes of volume can be brought about either by moving the tongue, or by opening the jaws more or less widely, and changes of opening by means of the lips. It will be seen therefore that we can make changes in the quality of the voice by altering the shape and size of the mouth, and it is by such changes that the different vowels are produced.

192. Vowel Theories. There are at present two theories as to the cause of the differences between the vowels, each of the theories having its adherents.

All are agreed that a vowel sound contains a long series of harmonics, some one of which is strengthened by the resonance of the mouth. In certain cases the mouth cavity is divided into two by the arch of the tongue, and in these cases two harmonics are strengthened, for each of the parts of the mouth cavity has its own natural period. The point at issue is whether the strengthened tone is fixed in pitch, whatever may be the pitch of the note on which the vowel is sung, or whether it moves up and down with the pitch of the note, always remaining at the same interval above the fundamental. The two theories are called the *fixed pitch* and the *relative pitch* theories respectively. Helmholtz believed the fixed pitch theory to be true.

QUALITY OF MUSICAL NOTES

Taking the vowel *o*, for instance, as in the word *note*, he found the strongest resonance was always at $b^1\flat$, whose vibration number is 466, whatever might be the pitch of the note to which *o* was sung. He gives the following scheme for the maximum resonance due to the mouth cavity when different vowels are sung. The vowels are to be given their German pronunciation.

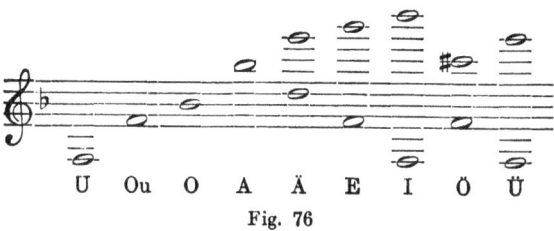

U Ou O A Ä E I Ö Ü

Fig. 76

According to the fixed pitch theory then, the mouth is set to a definite shape for each vowel, and retains that shape unchanged when a scale is sung. This theory is more generally held than the relative pitch theory at the present time—at least as giving the main cause of the difference between the vowels.

The relative pitch theory states that for a given vowel the harmonic which is strengthened by the mouth resonance is a certain definite member of the harmonic series, and so moves up and down with the fundamental. This view assimilates the vowel characteristic to what we have called the quality of the notes of a musical instrument. According to this theory the quality of the vowel is the same for all pitches of the fundamental, as the relative strength of the harmonic constituents remains the same. Hence if we sing a rising scale on some one vowel, the mouth must be altered at each step of the scale, so that its resonance pitch may rise, as the pitch of the fundamental rises.

We shall have more to say on the vowel theories when we have described the Phonograph.

193. Harmonic constituents of the notes of Musical Instruments. We shall conclude the chapter with a brief account of the distribution of the harmonics in the notes of a few instruments.

The Pianoforte has the second harmonic nearly as strong as the fundamental. The third, fourth, etc. fall off rapidly in intensity, until above the sixth or seventh the harmonics are very faint.

The Violin has a long series of harmonics falling off gradually in intensity. They fall off somewhat rapidly as far as the fourth, and then more slowly.

A Stopped Organ pipe has only the odd harmonics. If the pipe is wide the note is practically pure; if it is narrow, the third harmonic is strong and the fifth is perceptible.

An Open Organ pipe has the full series so far as they extend. A wide pipe has the octave fairly strong, and one or two others perceptible. A narrow pipe has a series of gradually diminishing intensity as far as the sixth or seventh.

The Flute gives a nearly pure tone. The octave is faintly audible, but no other harmonics can be heard.

The Clarinet has the third, fifth and seventh harmonics fairly strong. The fourth, sixth and eighth are also very faintly audible.

The Hautboy, the Brass Instruments, and the Human Voice are alike in having a full series of harmonics falling off gradually in intensity, but they differ in the height to which the series extends. The note of the French Horn has no appreciable harmonics above the sixth, the Trumpet and Trombone have harmonics quite perceptible as far as the eighth or ninth, whilst the Hautboy and the Human Voice extend to the sixteenth or higher before becoming inaudible.

CHAPTER X

ORGAN PIPES

Organ pipes may be divided into two main classes, Flute or Flue Pipes and Reed Pipes, but within each of these classes there are many varieties differing in shape and material.

194. Action of the mouth of a Flue Pipe. Fig. 77 is a section of a typical Flue Pipe. Air is blown into the pipe at A. It issues from the long narrow slit B in a thin sheet, which strikes the sharp edge C of the front of the pipe, and sets up vibrations in the body D. The pipe may be made of metal or of wood, its section may be circular or square, and its upper end may be open or closed.

The manner in which the sheet of air excites vibrations in the pipe is uncertain, but it is probable that something of the following kind takes place. We shall assume that the pipe is closed at the upper end as in the figure.

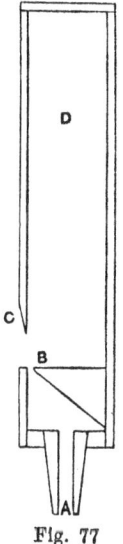

Fig. 77

The sheet normally strikes the edge C, but some accidental circumstance, such as a slight movement of the air, may deflect it inwards. A puff of air then enters the pipe and causes a condensation, which travels to the closed upper end, is reflected there, and returns to the mouth. The rise of pressure resulting from the arrival of the condensation at the mouth deflects the sheet of air outwards. As the mouth is to be regarded as an open end, the condensation is now converted by reflection into a rarefaction, and the amount of rarefaction is increased by the action of the sheet of air, which is now blowing across the outside of the mouth, and by a well-known hydrodynamical principle lowers the pressure inside the pipe*.

The rarefaction then travels to the top, and down again to the mouth. On its arrival at the mouth it sucks, so to speak, the sheet of air inwards, a condensation is produced, and the whole process is repeated. Thus, if the vibration is once started, it will be maintained by the oscillations of the sheet of air, for, whenever a condensation arrives at the mouth and is converted by reflection into a rarefaction, the conversion is helped by the sheet passing outside the pipe, and similarly the change from a rarefaction to a condensation is helped by the sheet passing inside the pipe.

195. Period of vibration of a closed pipe. The account we have given of the action of the blast serves also to shew that the period of vibration of a closed pipe is the time taken by a pulse in travelling twice up and down the pipe, for a pulse starting as a condensation makes two journeys to the top and back before starting again as a condensation, or, it is only after two journeys that it completes its cycle.

The distance a wave travels in one period is one wave-length, and therefore the wave-length of the tone given out by a closed pipe vibrating in the manner described is four times the length of the pipe. We shall see presently that the pipe can vibrate in other modes, which give a different relation between the wave-length and the length of the pipe. The mode we have described is that corresponding to the lowest or fundamental tone of the pipe.

196. Period of vibration of an open pipe. Let us consider next the case of a pipe which is open at the top as well as at the mouth. A condensation started at the mouth travels to the open top, and is returned thence as a rarefaction. On arriving at the mouth it again changes phase, and starts its second journey up the pipe as a condensation. Thus it always starts from the mouth in the same phase, and therefore

* This phenomenon can be shewn by holding a glass tube with its lower end in water, and blowing strongly across its upper end. The water will rise a little in the tube, shewing that the pressure has been reduced by the current of air.

the time it takes to travel from the mouth to the top and back is one period, and the wave-length is twice the length of the pipe.

If then we have two pipes of the same length, one open and the other closed, the open pipe will give a tone whose wave-length is half that of the tone given by the closed pipe, or the open pipe will sound an octave higher than the closed pipe.

197. Correction for an open end. This result is not strictly true. We saw in a former chapter that the reflection from an open end must be regarded as taking place a little way beyond the end, and consequently the open pipe, having two open ends, must be regarded as being acoustically a little longer than the closed pipe, which has only one open end. Twice the length of the open pipe with its two additions is more than one half of four times the length of the closed pipe with its one addition, and therefore the open pipe is rather less than an octave above the closed pipe. Blow an open organ pipe and then cover its end. The note will be heard to fall a little less than an octave.

The addition that has to be made to the length of a pipe with an open end is called the *correction for the open end*. The smaller the bore of the pipe is in proportion to the wave-length, the smaller is the correction. For the present we shall suppose the pipe is narrow and not very short, and shall neglect the correction.

198. Overtones of flue pipes. Another method of deducing the mode of vibrations of pipes may usefully be given, as it leads more directly to the relations between the overtones. We may regard the blast at the mouth as sending a train of waves along the pipe. These waves are reflected from the other end—open or closed, as the case may be—and by the superposition of the reflected train on the direct train stationary vibrations are produced in the pipe, in the manner explained in Chapter IV.

We may then imagine the pipe to enclose any number of segments of a train of stationary waves, subject to two conditions :—

(1) An open end must be situated at an antinode, for the change of pressure must be a minimum at an open end.

(2) A closed end must be at a node, for there cannot be any motion at a closed end.

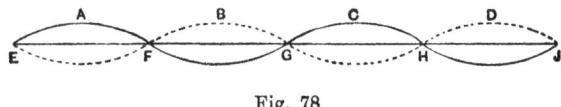

Fig. 78

Let Fig. 78 represent the displacement curve for a series of stationary waves at two instants. The full curve shews the maximum displacements of the various particles of air in one direction, and the dotted curve shews their maximum displacements in the opposite direction half a period later.

The nodes are situated at E, F, G, H and J and the antinodes at A, B, C and D, and any two consecutive nodes, or any two consecutive antinodes, are half a wave-length apart. As an open pipe must have an antinode at each end, we may regard it as enclosing the length AB, or AC, or AD of the train. In the first case it encloses half a wave, in the second case a whole wave, in the third case three half waves, and so on.

199. Overtones of an open pipe. If we are considering the various modes of vibration of an open pipe of fixed length, we must imagine the wave-length of the train to shrink step by step, so as to allow 1, 2, 3, etc. half waves to fill the pipe, always beginning and ending at an antinode.

We can now find the wavelengths and modes of vibration of the various possible tones of the pipe.

The fundamental may be represented diagrammatically by No. 1 of Fig. 79. There is a node in the middle of the pipe, and an antinode at each end, and the wave-length is double the length of the pipe, as we saw before.

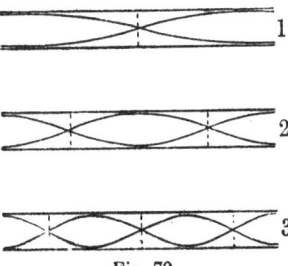

Fig. 79

Admitting that we have stationary vibrations in the pipe, and that the ends are antinodes, it is clear that we cannot have any mode of vibration that will give a lower note than this, for there must be at least one node between any two antinodes. If we have more than one, as in No. 2 or No. 3, the wave-length must be less, and the pitch higher.

The mode of vibration that gives the first overtone is shewn in No. 2. Here there are two nodes, each one quarter of the length of the pipe from an end. The pipe contains one wave, and therefore the wave-length is one half, and the frequency twice what it was for the fundamental.

Similarly the third mode has three nodes, one of which is in the middle and each of the others one-sixth of the length of the pipe from an end. The wave-length is one-third and the frequency three times that of the fundamental.

Evidently this process can be continued indefinitely, and we conclude that an open pipe can give a series of tones whose frequencies are proportional to the series of numbers 1, 2, 3, 4, etc.— the same series of tones as we found for a stretched string.

200. Overtones of a closed pipe. The case is different for a closed pipe. Here the closed end is a node and the open end an antinode. The lowest note for which these conditions are fulfilled is one for which one quarter wave-length fills the pipe as in No. 1.

The next higher note possible is one which takes in the part from E to B of Fig. 78. This mode is shewn in No. 2 of Fig. 80. There is a node at Q. PQ is half a wave-length and QR is a quarter of a wave-length, and therefore the node Q is one-third of the length of the pipe from the open end. No. 1 contains one quarter-wave and No. 2 contains three quarter-waves; and therefore the second mode has three times the frequency of the first.

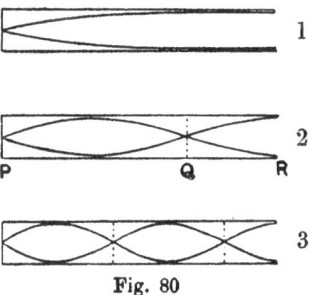

Fig. 80

The third mode has two nodes in addition to that at the closed end. One of these nodes is one-fifth the length of the pipe from the open end. The pipe contains five quarter-waves and the frequency is five times that of the fundamental.

This process again can be continued indefinitely, and thus we see that the possible tones of a closed pipe have frequencies proportional to the series of odd numbers 1, 3, 5, etc.

201. Experiments on the overtones of pipes. An open pipe then can give the full series of harmonics shewn in Fig. 31 whilst a closed pipe gives only every alternate tone beginning with the lowest.

An ordinary tin whistle is merely an open organ pipe, whose length can be altered by uncovering the holes in turn. If all the holes are kept covered, the first three or four tones of the series can be produced without much difficulty. Blow gently, and the fundamental is heard. Blow harder, and the note jumps an octave to the first overtone. Blow still harder, and the note goes up to a twelfth above the fundamental.

To shew the overtones of a closed pipe a small closed organ pipe may be used. The first overtone, a twelfth above the fundamental, is easily produced by vigorous blowing. The next is two octaves and a major third above the fundamental. It will probably be found difficult to blow the pipe hard enough to produce this note. It is easier to produce the overtones on a narrow pipe than on a wide one.

The manner in which the air vibrates in a pipe can be illustrated by a spiral spring similar to that described in Chapter IV. Hang such a spring vertically, draw down its lower end slowly, and then release it. The coils of the spring will execute vibrations similar to the vibrations of the successive layers of air in a closed pipe giving its fundamental. The upper end of the spring corresponds to the closed end of the pipe where there is no motion, and the free end corresponds to the open end of the pipe. Any one coil of the spring describes harmonic vibrations, the amplitude of the coil situated at the free end being the greatest, and that of the coil at the fixed end being zero. At any moment all the coils of the spring are moving in the same direction, either closing in on the fixed end, or spreading outwards from it, and the velocity of a coil is greater, the farther that coil is from the fixed end.

Any other mode of vibration of a closed or open pipe can be illustrated by imagining two or more such springs joined end to end. For an open pipe giving its fundamental imagine two such springs with their fixed ends joined together, and the phases of the vibrations so adjusted that each of the two halves has its greatest extension at the same moment. In this case the state of condensation or rarefaction is the same at any moment at points equidistant from the centre of the pipe. The student will find it a useful exercise to work out one or two other cases in the same way.

The positions of the nodes and loops in an open pipe can be found experimentally by lowering into an organ pipe with one side made of glass a small paper drum, on which a pinch of sand is sprinkled. When the drum is anywhere near an antinode, the motion of the air makes the sand dance, whilst when it is at a node, the sand does not move. By this means it can be shewn that, when the pipe is sounding its fundamental, there is a node at the middle point, and when it is overblown so as to sound its first overtone, there are two nodes, each a quarter of the length of the pipe from an end. It will be seen also that in all cases the sand is violently agitated when the drum is near either end of the pipe.

202. The Manometric Capsule. König's Manometric Capsule may also be used for shewing the positions of the nodes.

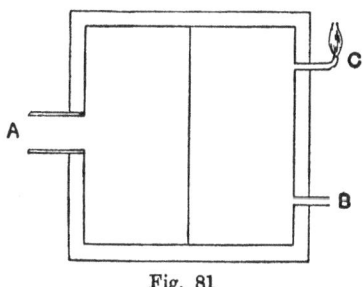

Fig. 81

The manometric capsule consists of a small box divided into two parts by a thin flexible membrane.. The space on one side of the membrane has a short tube A, which can be inserted in the wall of the pipe. The space on the other side of the membrane has two openings. Through the lower opening B gas is admitted, and the upper opening is provided with a pinhole burner, where the gas burns with a small flame.

If the tube A is inserted near a node of the pipe or, as is more usually the case, the left-hand half of the capsule is omitted, and the diaphragm forms part of the wall of the pipe, the variations of pressure will make the membrane vibrate, and the flame will rise and fall in time with the vibrations. The dancing of the flame is too rapid to be detected by the unaided eye, but is easily seen with the help of a revolving mirror. A cubical box has mirrors on its four vertical faces, and is made to rotate about a vertical axis. The image of the flame seen in the mirror travels so quickly across the field of view, that it appears as a band of light. If the flame is at rest the band is continuous; if it is dancing, the upper edge of the band shews a series of teeth.

If the capsule is placed at the middle of an open organ pipe, which is giving out its fundamental, the band will shew teeth, for we have seen that there is a node at the middle of the pipe. If now the pipe is blown more strongly so as to give out its next higher tone, the teeth will almost disappear, for the capsule is now at an antinode. The teeth will not disappear completely, for there are always harmonics present, some of which have nodes at or near the middle of the pipe.

203. Methods of tuning flue pipes. There are several methods of tuning open pipes. One method is to make a hole in the side of the pipe near the open end and cover this with a flap of lead fixed to the pipe by one edge. If the flap is pulled away slightly from the pipe, the pipe is virtually shortened, and the pitch is raised.

Another method is to make the upper part of the pipe slide telescopically over the lower part, so that the length can be altered.

Many of the flue pipes in an organ consist of cylindrical

tubes of soft metal. These can be tuned by altering slightly
the size of the opening at the top of the pipe. A hollow cone
of wood pressed over the end closes the opening a little and
lowers the pitch. A solid cone pressed into the opening
enlarges it and raises the pitch.

Metal pipes often have two flaps standing out, one on each
side of the mouth. If these are bent a little towards each
other, the pitch is lowered, and vice versa.

The principle of these last two methods will be understood,
if the pipe be regarded as a resonator, for we have seen that
enlarging the mouth of a resonator raises its pitch.

Stopped pipes generally have the upper end closed by
a tightly fitting plug, and a pipe can be tuned by pushing the
plug in or pulling it out, so as to alter the length.

204. Determination of the correction for the open end. We must return now to the correction for the
open end of a pipe, and explain in the first place how it has
been determined, and afterwards how its existence affects the
pitches of the proper tones of the pipe.

The theoretical determination of the amount that must be
added to the open end of a pipe to give the effective length
has hitherto proved to be too difficult for mathematicians,
except in the case of a narrow cylindrical pipe with an infinitely
large flange at the end. In this case the correction is found
to be $\cdot 82\,R$, when R, the radius of the pipe, is small compared
with the wave-length.

Unfortunately flanged pipes are of little practical importance, and we are compelled to rely on experiment for finding
the value of the correction for such pipes as are actually used
in musical instruments.

Rayleigh determined the correction for an unflanged pipe
by finding the change in the pitch when the flange was removed.
Two pipes of nearly the same pitch were blown together, one
of the pipes having a flange which could be removed. The
number of beats per second was counted, first when the flange
was in position, and next when it was removed. The difference
between the numbers of beats per second in the two cases is
the change in frequency caused by the removal of the flange.

It is clear from what was said of the mouth of resonators in Chapter VIII that the flange must act in the direction of hindering the free egress of air from the pipe, as it confines the stream lines into a smaller space. Hence removing the flange has the same effect as enlarging the mouth of a resonator, that is to say, it raises the pitch of the pipe.

The pipe used by Rayleigh had a frequency 242, and it appeared that the effect of the flange was to reduce the frequency by $1\frac{1}{2}$. Rayleigh took the velocity of sound at 60° F. to be 1123 ft. per sec.; the effective length of the pipe was therefore about 28 inches. The radius was 1 inch. Thus the correction due to the flange is the same fraction of 28 in. as $1\frac{1}{2}$ is of 242, or about $\cdot 2\,R$. Since the correction for a flanged end is known to be $\cdot 82\,R$, that for the open end of an unflanged pipe is $\cdot 62\,R$.

Blaikley determined the correction by immersing the lower part of a thin brass tube in water, and finding the length of the unimmersed part when the tube resounded most strongly to a fork of known pitch. The water surface forms a closed end to the pipe, and therefore the pipe gives maximum resonance when its length with the correction added is $\frac{\lambda}{4}$, $3\frac{\lambda}{4}$, $5\frac{\lambda}{4}$, etc.

Blaikley measured the two shortest lengths of pipe which gave resonance. Call these l_1 and l_2, then $l_2 - l_1 = \frac{\lambda}{2}$, for no correction is to be added to the half-wave from the bottom of the pipe to the node in the second mode of vibration.

Also $l_1 + c = \frac{1}{4}\lambda$,
∴ $l_2 - l_1 = 2\,(l_1 + c)$
or $c = \dfrac{l_2 - 3l_1}{2}$.

Blaikley found $\cdot 58\,R$ as the mean value of the correction.

205. Effect of the correction for the open end on the overtones. If the correction depended only on the

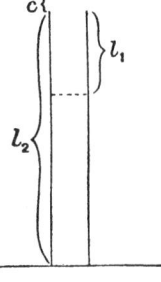

Fig. 82

width of the pipe, and not on the wave-length of the note produced, its existence would not affect the relative pitches of the natural tones of a pipe. The pipe would merely have to be regarded as longer by an amount c, and what has been said as to the pitches of the natural tones would still hold, each tone being a little flatter than it would be if no correction were needed.

This is not generally the case, for the correction varies with the wave-length. It is found that with open pipes of wide bore the overtones are all sharper than the harmonics of the fundamental, and the divergence is greater for the overtones of higher order. The effect is less marked with narrow pipes, and for these the lower proper tones are fairly concordant with the harmonic series.

206. Effect of the correction for the open end on the quality. The existence of the correction for the open end has an effect on the quality of organ pipes. The pipe may be regarded as a resonator, which is excited by the vibrations of the sheet of air at its mouth. These vibrations are maintained unchanged whilst the pipe is sounding, and are therefore periodic, and can be expressed by the harmonic series. It has been stated that a resonator will respond appreciably to a periodic force only when there is a harmonic constituent of the force having a period near that of one of the proper modes of vibration of the resonator. A narrow pipe has proper tones which are nearly in accordance with the harmonic series, and therefore fall near the constituents of the force. For such a pipe therefore the resonance will extend to a considerable number of the lower members of its modes of vibration, and the note given out by the pipe will contain many harmonics. It should be noted that the constituents of the note of the pipe are *harmonics* and not the proper tones of the pipe. If a particular proper mode of vibration has a period near a harmonic of the mouth vibration, the vibration excited is not in this natural period of the pipe but is "forced" into the period of the harmonic. The nearer the proper tone and the harmonic are to each other in pitch, the greater will be the intensity of the corresponding harmonic constituent of the note.

Wide pipes have proper tones which depart markedly

from the harmonic series. Such pipes therefore give notes which contain few harmonics, and these harmonics are low in the series. If a wide closed pipe is blown gently, it gives a note which is almost a pure tone.

The note of a narrow pipe is rich and full, that of a wide pipe is smooth and rather dull.

207. Conical Pipes. All the pipes of which we have spoken up to this point have been assumed to be of uniform bore throughout their length. Conical pipes are used to some extent in musical instruments, and should be mentioned here. Their theory is less simple than that of cylindrical pipes.

It has been shewn by Helmholtz that a pipe consisting of a cone closed at the narrow end has a series of proper tones whose frequencies are in the ratios 1, 2, 3, etc., or a closed conical pipe has the same series of proper tones as an open cylindrical pipe.

The hautboy, bassoon and clarinet all consist of tubes with a reed at one end, and, as we shall see later, the reed end is to be regarded as a closed end. The hautboy and bassoon have conical tubes with the reed at the narrow end and therefore rise an octave when overblown, whilst the clarinet has a cylindrical tube and therefore rises a twelfth.

208. Reed Pipes. We must now pass to the second group of organ pipes—the Reed Pipes. In these the vibration is caused by a tongue of metal, which vibrates in front of an opening in a metal plate, thus allowing a series of puffs of air to pass through the opening.

Fig. 83 is a section of one form of reed pipe. Air from the bellows of the organ is introduced into the box B through the tube A. The pipe D is fixed with its lower end inside the box. This end consists of a metal tube with one side flattened, and pierced with a long narrow opening at C. Over this opening is a rectangular tongue of

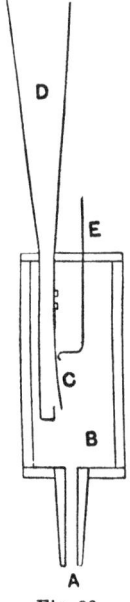

Fig. 83

metal called the reed, fixed at its upper end, but free for the rest of its length. The tongue is a little larger than the opening, so that when pressed down it cuts off completely the supply of air to the pipe. It is not quite flat, but bent outwards a little in a curve, and so, when it is at rest, air can pass from the box to the pipe. The curvature of the reed also makes the closing less sudden—the reed rolling over the hole, so to speak, and closing it gradually.

209. Free reeds and beating reeds. A reed such as that described in § 208 is called a *beating reed*, since at each vibration it strikes the plate in which the opening is cut.

There is another kind of reed called a *free reed*, which is slightly smaller than the opening, and so does not cut off the supply of air so completely at each vibration as does the beating reed.

Fig. 84 shews the two forms of reed. The full line is the vibrating tongue and the dotted line is the opening.

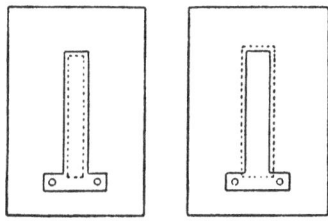

Fig. 84

In the cases where the reeds are associated with pipes, as in the organ, free reeds have been superseded by beating reeds almost completely. Many modern organs have no pipes with free reeds. When however the reeds are not associated with pipes, as in the harmonium and concertina, free reeds are used, since beating reeds give too harsh a tone, if they are not mellowed by the addition of pipes.

210. Action of the reed. The explanation of the action of a reed pipe is somewhat similar to that of a flue pipe, where the sheet of air might be regarded as resembling a reed in some respects, though, as we shall see, there is the important difference that the mouth of a flue pipe is an open end, whereas a reed pipe must be regarded as closed at the reed end

When a reed pipe is blown, air passes the sides of the reed into the opening. This causes a condensation to run up to the open end of the pipe, and return as a rarefaction. Meanwhile the reed has been closed by the rush of air, and is held closed by the arrival of the rarefaction. Consequently the pulse is still a rarefaction when it starts on its second journey. It travels to the open end, changes sign, and returns to the reed end as a condensation, where its pressure added to the elasticity of the metal tongue causes the reed to open, and the whole process is repeated. Thus we see that the period of vibration is the time taken by the pulse in travelling four times the length of the pipe, or the wave-length is four times the length of the pipe.

We have used this somewhat artificial explanation once more, because it shews why the reed end is to be regarded as a closed end, in spite of the fact that it is at this end that the air and the energy enter the pipe.

The reed is always closed when it is at a centre of rarefaction, and to a great extent *because* it is at a centre of rarefaction. It is always open when it is at a centre of compression, but it does not then behave as an open end, for the moment it opens it admits a stream of air at high pressure, which increases the condensation already existing.

211. Effect of the reed on the pitch of the pipe. When the reed is very flexible, like the thin cane reed of a clarinet, the note produced depends solely on the length of the pipe, the reed being constrained to vibrate in unison with the pipe. If the reed were sufficiently stiff and heavy, it would force the vibrations of the pipe, and the pitch of the note would depend solely on the natural period of the reed. The reeds used in organ pipes are of thin metal, and there is reciprocal constraint. The reed forces the vibrations of the pipe, and the pipe modifies the natural period of the reed. It is necessary in practice that the reed and the pipe should have nearly the same pitch, for, if this were not the case, the note would have a poor quality, and the pipe would not "speak" readily; that is to say, it would not begin to sound the moment the air was admitted.

212. Method of tuning a reed pipe. A reed pipe is tuned, not by altering the length of the pipe, but by altering the free period of the reed. A wire E passes through the top of the box B (Fig. 83), and its bent end presses against the reed, so that the part of the reed from its upper end to the place where it is touched by the wire is held against the seating, and only the part below the wire can vibrate. Hence, if the wire is pulled up, the reed is lengthened, and the pitch lowered, and vice versa. The reed is able to control the pipe within the range required for ordinary tuning, but if it is shortened by a considerable amount, the note jumps to a pitch somewhere near one of the higher proper tones of the pipe.

The reeds could, if desired, be used without pipes, and, as we have said, free reeds are so used in the harmonium. The chief reason for adding the pipe is that the tone is thereby greatly improved and strengthened. A beating reed alone gives a note of harsh and unpleasant quality from the strength of its high harmonics. When a pipe is added, such harmonics of the reed as fall near proper tones of the pipe are strengthened, and the others are unaffected. If the pipe is not very narrow, it is mainly the lower harmonics that are strengthened, and the note is therefore made less piercing and disagreeable.

The higher harmonics are always present to a greater or less extent, and are the cause of the characteristic penetrating quality of reed pipes.

213. Effect of temperature on Organ Pipes. Flue Pipes and Reed Pipes are both altered in pitch by change of temperature, but to a different extent.

Rise of temperature affects a flue pipe in two ways. The pipe gets a little larger from the expansion of the wood or metal of which it is made, and the density of the air within it is diminished. The increase of size of the pipe is so small as to have no appreciable effect on the pitch, but the change in the density of the air alters the pitch by an amount that is quite perceptible.

We saw that the velocity of sound in a gas is proportional to the square root of the absolute temperature, and the frequency of the vibrations of a pipe is v/λ, where λ is constant, being twice or four times the length of the pipe according as it is open or stopped. The frequency of the pipe is therefore proportional to the square root of the absolute temperature.

Let us suppose the temperature of the organ rises from 60° F. to 75° F. in the course of a performance. If then a flue pipe has a frequency 512 at 60°, it will have a frequency $\dfrac{\sqrt{459+75}}{\sqrt{459+60}}$ 512, or 519, at 75°. This is nearly a quarter of a semitone sharper.

The pitch of a reed pipe is dependent on the pitch of the reed as well as on that of the pipe, and these are altered in opposite directions by rise of temperature. The reed becomes a little softer and therefore flatter, whilst the pipe becomes sharper. The effect on the reed is in general less than that on the pipe, and therefore the pitch on the whole rises, though to a less extent than the pitch of the flue pipes. It follows that the different sections of an organ can be accurately in tune with each other only at one temperature

CHAPTER XI

RODS, PLATES AND BELLS

We shall conclude the discussion of the vibrations of different classes of bodies with a brief account of the modes of vibration in several miscellaneous cases. Some of these have only a theoretical interest, and nearly all require advanced mathematics for their complete investigation.

214. Longitudinal vibrations of Rods. Elastic rods can vibrate in many different ways. The vibrations may be longitudinal, transverse, or torsional; the ends of the rods may be fixed, supported, or free; and the rods may be of any section. Only a few of the possible modes need be mentioned.

The longitudinal vibrations of a rod are closely analogous to those of the column of air in an organ pipe. The fixed end of the rod corresponds to the closed end of a pipe or to a node, and the free end corresponds to an open end of a pipe or to an antinode. We may have waves of compression and rarefaction running along the rod, and stationary vibrations produced by their reflection at the end, exactly as we had in the case of a column of air. The chief difference between the two cases is that the rod needs no correction corresponding to the correction for the open end of the pipe.

The velocity of waves in a rod is expressed by the usual formula $\sqrt{\dfrac{E}{D}}$, where E is the elastic constant and D the inertia term involved in the particular kind of wave that is being considered. In the case of longitudinal waves E is Young's Modulus, and D is the density of the material of which the rod is made. The formula shews that the velocity is independent of the tension in the cases where the rod is fixed at both ends and stretched. It is also independent of the thickness, provided the material remains the same. Doubling the cross section doubles the inertia of any layer, but it also doubles the forces of restitution, and so leaves the velocity unchanged.

215. Longitudinal vibrations of a rod fixed at both ends. The longitudinal vibrations in a rod fixed at both ends are not analogous to the vibrations of the *complete* column of air in any organ pipe, for a pipe cannot have both ends closed. They resemble the vibrations of the column bounded by two nodes in one of the higher modes of a pipe. They are easily produced in a wire 6 or 8 feet long, firmly fixed at its two ends, and provided with some appliance by which it can be drawn tight. If such a wire is rubbed near its middle with a piece of leather dusted with resin, it will give its fundamental tone.

If the wire is held at its middle point between the finger and thumb, and one of the halves is rubbed, the note produced will be an octave higher than before.

It is needless to enter into details of the various possible modes of vibration. The fixed ends are always nodes, and there may be any number of nodes between them. It follows as in the case of the transverse vibrations of a string that the natural tones must form the full harmonic series.

216. Longitudinal vibrations of a rod fixed at one end. A rod fixed at one end has modes of vibration similar to those of the air in a closed pipe. The fixed end of the rod is always a node, and the free end an antinode. The natural tones of the rod have frequencies proportional to the odd numbers 1, 3, 5, etc., and the positions of the node for each tone is the same as for the corresponding tone of a closed pipe. The tones are easily produced by rubbing with resined leather a thin metal rod clamped in a vice at one end.

217. Longitudinal vibrations of a rod free at both ends. The longitudinal vibrations of a rod free at both ends are analogous to the vibrations in an open organ pipe. They have a special interest from their being the kind of vibrations used in Kundt's apparatus to be described in Chapter XII. Hold a glass rod or tube five to six feet long at its middle, and rub it lengthways near the end with a piece of wet flannel. The fundamental tone of the rod will be produced without much difficulty. The point held in the hand is a node, and the two free ends are antinodes, and therefore the wave-length in the glass is twice the length of the rod. In order to produce the first overtone hold the rod one quarter of its length from one end, and rub the shorter section. We have now two nodes, each a quarter of the rod's length from an end, and the note produced is an octave higher than the fundamental. It is theoretically possible to produce the full series of harmonics, but difficult in practice to get anything above the second with a rod of reasonable length. The tones produced are very powerful. They are occasionally so powerful as to shatter the glass.

The student should avoid making the mistake of supposing that the wave-length *in air* is, for instance, four times the length of the rod, when a rod fixed at one end gives out its fundamental. When we spoke of organ pipes we assumed that the vibrating substance was the same inside and outside the pipe, and the wave-length was the same inside and outside. With rods the case is different. The *period* of vibration of the rod is the same as the period of the vibrations produced by it in the air, but the wave-length is different. It is evident from the equation $v\tau = \lambda$ that, if τ is the same for the rod and the air, the wave-length in the rod is to the wave-length in air, as the velocity of waves in the rod is to their velocity in air. The velocity of sound in an iron rod is about 15 times as great as in air, and therefore an iron rod 4 ft. long fixed at one end would produce air waves a little over a foot in length.

218. Transverse vibrations of rods. There are many modes in which a rod can vibrate transversely, according to the way in which it is supported. It may have one or both ends free, one or both ends firmly fixed, or both ends resting on supports. When both ends are free it may be supported or fixed at one or more intermediate points.

If both ends are supported, the rod takes a form such as No. 1 of Fig. 85, where it can change its direction at the ends*.

If both ends are fixed, it takes the form of No. 2, where the direction of the ends cannot change. The elastic forces resisting bending are clearly greater in the second case, and therefore the pitch is higher, if the two rods are of the same length and thickness.

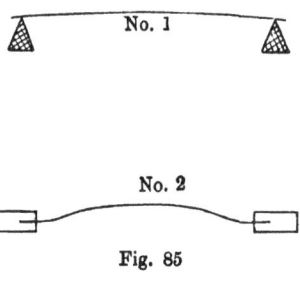

Fig. 85

Only two cases of transverse vibrations need be considered, namely, that in which one end is fixed and the other free, and that in which both ends are free, the rod being supported at two intermediate nodes. The former, which we may term a fixed-free vibration, is the more important.

219. Transverse vibrations of a rod fixed at one end.
The tongues of reeds, the vibrators of musical boxes, and the prongs of tuning forks may be regarded as fixed-free rods.

The first three modes of vibration of a fixed-free rod are shewn in Fig. 86, where A, B, and C are nodes. The modes of vibration bear some resemblance to those of a closed pipe, but there are important differences. The natural tones of a closed pipe have frequencies in the ratio $1:3:5:$ etc., whilst those of a fixed-free rod do not follow the harmonic series, but rise in pitch much more rapidly. The

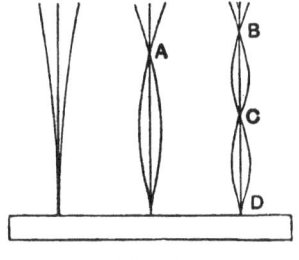

Fig. 86

* The rod could not be made to vibrate with its ends supported as shewn in the figure unless the amplitude were small. If it were large the rod would jump off its supports.

relative frequencies for the first three modes are approximately 1, 6¼, 17½.

The nodes too are not in the same positions as in a closed pipe. The node in the second mode is a little more than a fifth of the length of the rod from the free end; those in the third mode are about one eighth and one half respectively. The length BC is not the same as CD, though the two segments have, of course, the same period. The reason is that at D the direction of the rod is fixed, whilst at B and C the direction can change. The greater length of CD compensates the greater elastic forces due to the constraint at D.

The vibrations are easily produced in a visible form in a stout brass or steel wire fixed in a vice at its lower end. The wire should be of such a length that the vibrations in the first mode are quite slow—say one a second—otherwise those in the higher modes will not be easily seen. In order to produce the second mode hold the wire between the finger and thumb about one fifth of the length from the top, and pluck below the hand. The vibrations will be seen to continue in this mode when the fingers are removed. The third mode can be produced in a similar way.

The frequencies of the fundamentals of rods of the same thickness and material but different lengths are inversely proportional to the squares of the lengths.

Rods of the same length and of rectangular section, the direction of vibration being parallel to one of the sides of the rectangle, have frequencies proportional to the length of the side of the rectangle in the plane of vibration, and independent of the dimensions of the rectangle at right angles to the direction of vibration. These relations are easily seen to be true, for if a rod is doubled in width in the direction at right angles to the plane of vibration, both the inertia and the elastic forces are doubled, and therefore the period is unaltered. If the thickness in the plane of vibration is doubled, the inertia is doubled but the elastic forces are increased eightfold, and therefore the period is halved.

220. Wheatstone's Kaleidophone. It will be seen then that a rod of rectangular section fixed at one end has different periods of vibration in two directions at right angles to each other. If the free end is drawn aside in a direction not parallel to either of the sides of the rectangle, and is then released, both vibrations will be executed simultaneously, and the motion of the free end will be that got by compounding the two. If the dimensions of the rod are such that the periods of the two vibrations bear a simple ratio to each other, the free end will describe one of Lissajous' Figures. Such a rod with a small bright bead fixed to the free end to permit the figure to be seen is known as Wheatstone's Kaleidophone.

The Kaleidophone as more commonly constructed consists of two thin strips of steel joined end to end, the plane of one of the strips being at right angles to that of the other. One end of the Kaleidophone so formed is clamped in a vice. Each strip then vibrates only in a direction at right angles to its own plane and the vibration of the free end is compounded of the vibrations due to the two parts separately. The advantage of this form of the instrument is that the period of vibration of the lower strip can be varied by clamping the strip at different points and so altering its length. It is thus possible to vary the ratio of the frequencies of the two component vibrations and to produce several of Lissajous' Figures with the same instrument.

221. The musical box. The vibrator of a musical box consists of a comb cut from a steel plate. The teeth of the comb form a series of fixed-free rods, and are graduated in length so as to give a musical scale. The teeth are set vibrating by small pins fixed in the surface of a revolving drum in such positions that they pluck the ends of the rods at the proper moments.

222. Method of tuning a rod. The pitch of a fixed-free rod can be adjusted by scraping. If a little is scraped off near the free end, the inertia is diminished, whilst little change is made in the elastic forces, and therefore the pitch is raised. If the rod is scraped near the fixed end, the elastic forces are weakened without much change in the inertia, as there is little motion at this part of the rod, and therefore the pitch is lowered. The free reeds of harmoniums are fixed-free rods, and they are tuned by scraping away the metal at the free end or the base, according as it is desired to raise or lower the pitch.

223. Transverse vibrations of a rod free at both ends.

A rod free at both ends takes the forms shewn in Fig. 87 when it gives out its fundamental. A and B are nodes, and the bar may be supported

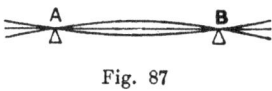

Fig. 87

at these points by wedges of some soft material such as india-rubber or felt without interference with the vibrations. Each node is ·22 l from an end, where l is the length of the rod.

The higher modes have 3, 4, 5, etc. nodes, and the frequencies of the vibrations in the various modes bear no simple relation to those of the harmonic series.

Bars free at both ends are used in the harmonicon. The bars consist of flat strips of glass, metal, or wood cut to such lengths as will give the musical scale. The lengths are inversely proportional to the square roots of the required frequencies. The strips are supported at their nodes on strings, or on strips of wood covered with felt, and are made to vibrate by being struck with a hammer.

224. The tuning-fork.

The tuning-fork may be regarded as two fixed-free rods on the same base. It is an instrument of the greatest value in acoustical experiments from its constancy as a standard of pitch. It is little affected by external conditions, and if kept free from rust, and not subjected to excessive ill treatment, it retains its pitch for years without appreciable change. The only correction needed is for change of temperature, and this is so small as to be negligible for ordinary purposes. A rise of temperature increases the size of the fork and diminishes its elasticity, but the latter change has much the greater effect on the pitch. It is found that the frequency of a fork diminishes by one ten thousandth of its amount for a rise of one Centigrade degree. Hence it would require a rise of 20° C. or 36° F. to lower the frequency from 512 to 511.

The overtones of a fork do not belong to the harmonic series. Their pitch depends on the shape of the fork, but they are always much higher than the fundamental. The first is 2 to 3 octaves, and the second 4 to 5 octaves above the fundamental. They are usually audible for only a few seconds after the fork is struck, as they die away more quickly than the fundamental.

A fork mounted on a resonance box gives a note that is practically a pure tone. The box, like the fork, has a series of natural tones, but in general none of the tones of the box lie near the tones of the fork except the fundamental, and therefore only the fundamental is strengthened.

A tuning-fork is sometimes regarded, not as two fixed-free rods, but as a single free-free rod bent at its middle. This second way of regarding the vibrations is useful, as it gives a clearer idea of the manner in which the vibrations in the fork cause vibrations in the resonance box. Fig. 88 shews three stages in the bending of a straight rod into a fork. There are two nodes in each case, those in the fork being at the base of the prongs. It will be seen on considering the successive stages that the short piece between the nodes at the base of the fork vibrates up and down, when the prongs vibrate outwards and inwards, and the up and down motion is communicated to the top of the resonance box, and so to the air.

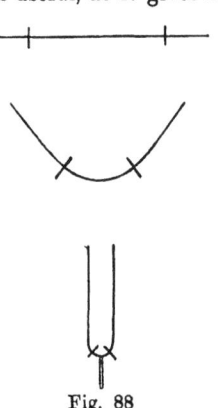

Fig. 88

The method of tuning a fork is similar to that used for tuning reeds. The prongs are filed near their free ends to raise the pitch, and near their bases to lower the pitch, as was explained in § 37.

225. The Triangle. The Triangle used in orchestras is a steel rod bent into a shape convenient for suspending and striking. It is struck with a steel rod, and its note contains a multitude of powerful constituents lying so close together that no definite pitch can be recognized. It is an extreme case of the quality of note that is called "metallic," if indeed it can be said to give a note.

226. Vibrations of Plates. Chladni's Figures.
The vibrations of flat plates have not much practical importance, but they are rendered interesting by the simplicity and beauty of the method devised by Chladni for shewing the nodal lines in the various modes of vibration.

The plates used for producing Chladni's Figures may be square or round, and may be fixed at any point We shall take as an illustration a square plate fixed at its centre. The plate is generally of glass or brass, 12 to 15 inches square, and is held firmly at its centre by a screw passing through a hole in the plate, or by a clamp. It should be of the same thickness everywhere, as otherwise the figures produced will be irregular. Most plates shew some irregularities, and these may be ascribed to irregularities in the thickness or elasticity of the plate.

Scatter a very little fine sand evenly over the plate, hold the point A at the centre of one side with the finger and thumb, and apply a violin bow near the corner B. The plate will then give out the lowest note it is capable of producing, and the sand will gather along two diameters parallel to the sides of the square.

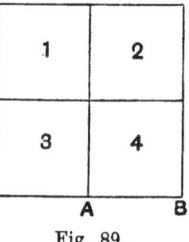

Fig. 89.

Some practice is needed to produce the notes easily. The bow should be held nearly vertical, and pressed fairly firmly against the plate. It should not be in contact with the plate at the moment of changing from an up stroke to a down stroke. In each stroke get the bow in motion before it touches the plate, and remove it before it comes to rest for the next stroke.

The lines of sand mark the lines on the plate which remain at rest, or the nodal lines. The quarters of the plate vibrate in such a way that when 1 is coming up, 2 is going down, 4 is coming up, and 3 is going down. It is a general rule in all Chladni's Figures that the sections divided from each other by any nodal line are always moving in opposite directions. The line clearly could not be a node, if this were not the case.

190   RODS, PLATES AND BELLS   [CH. XI

In the case we are considering 1 and 4 move together, and 2 and 3 move together, but in the opposite direction to 1 and 4. If then 1 and 4 are at any moment moving upwards and sending out a condensation, 2 and 3 must be moving downwards and sending out a rarefaction. The waves from the different quarters must therefore interfere to some extent. This can be proved by holding a card over each of 2 and 3 whilst the plate is vibrating, when the sound will be heard to be strengthened.

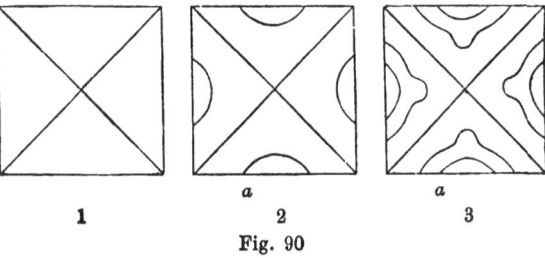

Fig. 90

The next higher note that can be produced gives the figure No. 1 of Fig. 90. Hold the plate at a corner and bow at the centre of one side to produce this figure. Nos. 2 and 3 are two other patterns that are easily produced. In each case bow at the centre of a side, and hold the plate at the point a where a nodal line meets that side.

Circular plates fixed at the centre give two classes of nodal lines. The first class consists of radial lines dividing the plate into an even number of sectors, and the second consists of circles concentric with the plate. The nodal circles are most easily produced, if the plate is not fixed at the centre, but supported on three small cones of wood placed so as to lie on a nodal circle, and is then tapped at the centre with a soft hammer. When there is only one nodal circle, it is about two-thirds of the radius from the centre.

227. Vibrations of Bells. Bells have modes of vibration somewhat similar to those of circular plates. The nodal lines are of two classes, radial lines running from the point of support of the bell to its edge, and dividing it into an even number of equal sections, and nodal circles running round the bell at various heights above the mouth.

Let us consider the mode in which there are four nodal lines from the summit to the rim and no nodal circles. This is the mode corresponding to the deepest tone that the bell can give out.

Fig. 91 represents the mouth of the bell when it is vibrating in this mode; A, B, C, D are the ends of the nodal lines. Between these points there are segments where the rim vibrates rapidly in and out. If at any moment it is outside its normal position at E and G, it will be inside at F and H. This is a similar statement to that made about Chladni's Figures. On crossing a node the direction of motion and the displacement are reversed.

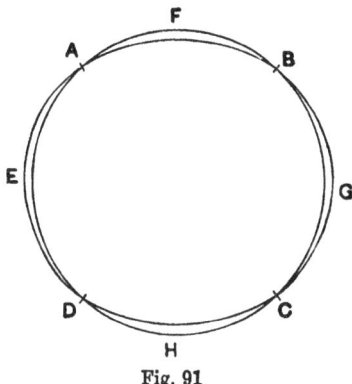

Fig. 91

It is evident that the blow of the clapper is suited to set up this form of vibration, but it can also be produced in another way.

When for a moment the bell has the shape in which the segment AB is inside its normal position the length of rim between A and B must be less than that between A and D; and a little later, when the segment AD is inside its normal position, and the segment AB is outside, the segment AB will be longer than AD. It follows that during each vibration a little of the rim is transferred through A from the segment AD to the segment AB and back again, or there is a tangential vibration of the rim at each of the nodes A, B, C, D. We might say that this tangential vibration has nodes at E, F, G, H and antinodes at A, B, C, D, but we must not treat it as a separate mode, for we cannot have it apart from the radial vibration. If we start the tangential vibration, it must of necessity be accompanied by the radial vibration.

The well-known method of making a wine glass sing by drawing a wet finger round its edge is an instance of this. The motion of the finger excites tangential vibrations, and these give rise to radial vibrations. This is easily seen by partly filling the glass with water, when the surface is seen to be disturbed at four points at equal distances from each other round the circumference. These points of greatest disturbance are the antinodes of the radial vibrations. They follow the finger round the glass, since the finger is always at a point of maximum tangential motion, and therefore always half way between two points of maximum disturbance of the water.

The note of a bell contains many and powerful constituents, which may or may not be harmonious with the main tone, and the art of the bell founder is largely devoted to finding by experiment the shape of bell which brings the more important constituents into harmonic relations with each other, as this conduces to sweetness of tone.

The best bells of the present day have the notes shewn in Fig. 92 as the most prominent constituents, assuming the note of the bell to be C.

Bells are an exception to the general rule that the lowest tone present is that which characterizes the pitch as a whole. The pitch of the bell whose main constituent tones are shewn in Fig 92 would not be taken to be the pitch of No. 1 but of No. 2. The tone No. 1 is called by bell founders the Hum Note. It

Fig. 92.

is more persistent than the others, and is often heard alone as the sound dies away. No. 3 is generally strong, and consequently a chord of a major third or a major tenth sounds bad when played on two bells. The second or the first tone of the bell of higher pitch is then a semitone above the third tone of the lower bell, and an unpleasant effect is produced by the clashing of the two tones with each other. A minor third or minor tenth is much better.

Bells are not generally accurately in tune when first cast. The maker puts them in tune and brings the subordinate tones into consonance with the main tone by placing the bells on a lathe and turning metal from the inside. Experience alone can shew from what part the metal must be taken to have the desired effect.

228. Vibrations of stretched membranes. A stretched membrane, such as the head of a drum, can vibrate in various modes somewhat similar to those of a plate. It is not necessary to enter into any details of the modes and nodal lines, as they are of no practical consequence. The great drum and side drums give mere noises, and are used only to accentuate the rhythm of the music. Orchestral drums have round their circumference a number of screws by which the parchment forming the drum head can be tightened and the pitch altered.

*229. The Method of Dimensions. We shall conclude the discussion of the modes of vibration of different classes of elastic bodies by giving the proof of an important law which holds for vibrations of any type.

If two elastic bodies made of the same material and performing vibrations of the same type have the same shape, and differ only in size, their periods of vibration will be proportional to their linear dimensions.

The theorem is most readily proved by what is known as the *method of dimensions*. Assuming that the proportions of the body remain the same whatever its size may be, the period of vibration may depend on any linear dimension l, the density of the material ρ, and the elastic coefficient of the material e. The period then can be expressed as a function of l, ρ, and e, and this function can in general be expanded into a series, which we may represent by $\Sigma A l^x \rho^y e^z$, where A is a numerical constant. The equation $\tau = \Sigma A l^x \rho^y e^z$ must be homogeneous; that is to say, every term must be of the same dimensions in mass, length and time. If this were not the case, a change in the size of the fundamental units as, for instance, the change from a foot to an inch as the unit of length, would render the two sides of the equation unequal. On one side of the equation we have only τ, which is a time, and has the dimensions T. On the other side l is a length and has dimensions L, and ρ is the mass of unit volume with dimensions $\dfrac{M}{L^3}$. There are two fundamental coefficients of elasticity, either or both of which may be involved according to the type of vibration. The coefficient of volume elasticity is, as we saw in Chapter V, of the form $\delta p \div \dfrac{\delta v}{v}$ and has therefore the dimensions of a pressure or $\dfrac{M}{LT^2}$. The coefficient of rigidity is the shearing force per unit of area which gives the unit angle of shear, or $\dfrac{\text{force}}{\text{area}} \div \text{angle}$, whence its dimensions are also $\dfrac{M}{LT^2}$. Hence we have in the expression for τ a series of terms with dimensions of the form $L^x \left(\dfrac{M}{L^3}\right)^y \left(\dfrac{M}{LT^2}\right)^z$ or $L^{x-3y-z} M^{y+z} T^{-2z}$, where x, y and z must be so related that the dimension of each term of the series is T. We have therefore, equating the indices of the units of length, mass and time respectively on the two sides of the equation,

$$0 = x - 3y - z,$$
$$0 = y + z,$$
$$1 = -2z,$$

which give $x=1$, $y=\tfrac{1}{2}$, $z=-\tfrac{1}{2}$ as the only possible set of values that will satisfy the conditions, and the series reduces to the single term $Al\sqrt{\dfrac{\rho}{e}}$. We see then that the period of vibration is proportional to the square root of the density of the body, and inversely proportional to the square root of the coefficient of elasticity—two conclusions which we reached by another method in Chapter II. We have also the result that the period is proportional to the linear dimensions. If we have, for instance, two tuning-forks which are geometrically similar in shape, and are made of the same material, but one is double the size of the other, the larger fork will be an octave lower than the smaller in pitch. Similarly if two spherical Helmholtz Resonators are so made that one has double the diameter of the other, and has also a mouth of double the diameter, the larger will resound to a note an octave lower than that to which the smaller resounds.

CHAPTER XII

ACOUSTICAL MEASUREMENTS

In the preceding chapters we have described a few such methods of measurement as fell naturally into the line of our argument. In the present chapter we shall gather together a number of other methods which have been used by investigators. The aim will be not so much to provide a complete list, as to describe representatives of the various types of measurement.

We shall deal first with the measurement of the pitch.

230. Measurement of pitch by Cagniard de la Tour's Siren. In Chapter I it was shewn that the frequency

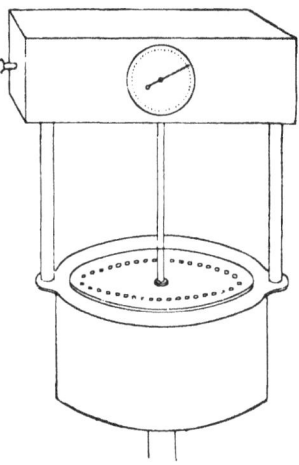

Fig. 93

of a note can be found by means of Savart's Toothed Wheel or the Disc Siren. These instruments are capable of giving only a rough approximation to the frequency. A better instrument of the same class, where the number of vibrations per second is counted directly, is Cagniard de la Tour's Siren shewn in Fig. 93.

This siren consists of two circular discs nearly in contact, of which the lower forms the fixed top of a wind chest, and the upper can rotate freely on a spindle. Each of the discs is pierced with a circle of holes, the holes in one disc corresponding in number and position with those in the other.

The wind chest is supplied with air from bellows with a pressure regulator. Suppose now the upper disc is rotating. Every time the holes in the two rows coincide a jet of air will escape from each hole in the upper disc, and if there are, for instance, 20 holes in each circle, there will be 20 puffs for each turn of the disc. Each puff will of course consist of 20 separate jets, but as they take place simultaneously they may be regarded as a single puff. The rotation is maintained by the pressure of the air in the wind chest, for the holes are cut obliquely as shewn in Fig. 94, so that the stream of air from the lower hole strikes the side of the upper hole. The greater the pressure of the air, the more rapid is the rotation of the upper plate, and therefore the higher the pitch of the note given out. The spindle of the upper plate is connected with a counter which can be put in or out of action as desired, and shews on a dial the number of rotations in a given time.

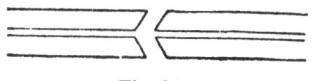

Fig. 94

In order to use the instrument for finding the frequency of the note of an organ pipe, for instance, start the siren by admitting wind from the bellows, and wait until its note ceases to rise. Then increase or diminish the pressure, until the siren and pipe are in unison. The final adjustment is most easily made by making the note of the siren a little too high, and bringing it down to that of the pipe by slightly reducing the air supply by means of a clamp on the supply tube. When the two notes have been brought to the same pitch, throw the counter into gear for a definite period, such as a minute, and find from the dial the number of revolutions in that period. This number multiplied by the number of holes in the circle gives the number of air vibrations per minute, or 60 times the frequency.

The determination cannot be made with any great accuracy for several reasons. Throwing in the counter reduces the speed of the disc a little, and puts the notes out of tune. This source of error can be avoided to some extent by leaving the counter in gear all the time, and when the tuning has been adjusted, taking the time of, say, 1000 revolutions by means of a stop watch.

It is not possible to keep the pitch of the siren quite constant during the experiment. The tuning is effected by making the beats between the two notes become gradually slower, until they disappear. If, as is generally the case, they reappear whilst the experiment is going on, they must be removed as quickly as possible, and this is not easy to do, for one cannot tell whether the siren is now too high or too low. The pitch can be lowered by touching the spindle gently with a piece of paper or a feather. If this quickens the beats, the pitch was already too low, and it must be raised by opening the clamp on the air tube a little. For further refinements of the method the reader is referred to Barton's Text Book of Sound.

231. Measurement of pitch by the Monochord.
We saw in Chapter III that the frequency of the vibrations of the string of a Monochord giving out its fundamental note is $\frac{1}{2l}\sqrt{\frac{T}{\rho}}$. If then the string is tuned to unison with a given tuning-fork, the frequency of the note of the fork can be calculated from this expression. An example of the calculation was given in Chapter III. The tuning is effected as usual by adjusting the length of the string until the beats disappear.

The ordinary form of the monochord is not suitable for measuring frequencies, for the friction of the pulley over which the string passes may cause the tension in the horizontal part to be different from that in the vertical part, and so not properly represented by the weight hung on the free end. It is better to fix the monochord in such a position that the whole string hangs vertically, and to hold the string against the bridge by pressing it with a thin piece of wood. The string must be thin, as otherwise its want of flexibility will render the formula inapplicable.

With these precautions frequencies can be determined by the monochord with an error of not more than about five vibrations per second.

232. Measurement of pitch by the Graphic Method. A method which gives much greater accuracy is that known as the Graphic Method. Many modifications of the method have been used, but we shall limit ourselves to the description of one of the simpler forms.

The method is an amplification of an experiment described in Chapter II. It was shewn there that when a tuning-fork with a light style attached to one of its prongs is drawn over a piece of smoked paper, or, what amounts to the same thing, the paper is drawn under the fork, a wavy line is traced on the paper. One wave-length of the curve is the distance the paper moves during one vibration of the fork. If then we know the velocity of the paper we can find the period and frequency of the fork by measuring the length of several waves.

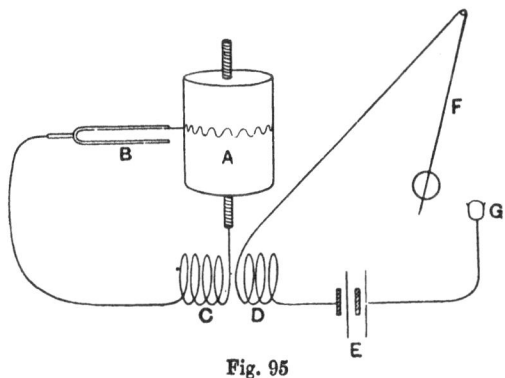

Fig. 95

It is evident that the experiment could not be carried out conveniently in this simple form. A better form is shewn in Fig. 95.

A is a cylindrical drum covered with smoked paper on which the fork *B* traces a curve by means of an aluminium style. The drum is rotated by clockwork, and is mounted on an axis with a screw cut on it. By this means overlapping of the various turns of the curve is prevented, since the drum travels endways as it rotates. *C* is the secondary of an induction coil, and has its ends connected to the fork and the drum respectively. The primary *D* of the same coil is connected to a battery *E*, and to a seconds pendulum *F*. The pendulum has below its bob a short platinum wire, which makes contact once a second with a mercury cup at *G*, and so allows a momentary current to pass through the primary. This gives rise to a momentary current of high electromotive force in the secondary circuit, and a spark passes from the style to the cylinder, making a mark on the paper.

If now the fork is vibrating and the cylinder rotating, we get a wavy curve, marked at intervals with spots which shew the ends of the seconds. It is only necessary then to count the number of waves in any interval of time measured by the spots on the paper, and divide by the number of seconds to find the frequency of the fork.

It will be seen that the result does not depend on the rate of revolution of the cylinder or on its regularity, but merely on the time of vibration of the pendulum. This pendulum is usually the pendulum of a good clock whose rate can be determined accurately by astronomical methods, and the method is therefore capable of a high degree of accuracy. It is to be observed however that the pitch determined is that of the fork with its style rubbing on the paper. The effect of the friction on the pitch is small, but it may introduce some slight error.

233. Measurement of pitch by the Tonometer.
We have seen that, when two notes have nearly the same pitch, the difference between their frequencies is equal to the number of beats per second, when the notes sound together. This fact was used by Scheibler in devising his Tonometer, which is very convenient for determining frequencies quickly.

The tonometer consists of a series of tuning-forks extending over an octave, the interval between any two consecutive forks in the series being such that they give about four beats per second. It is found that the difference between the frequencies of two notes can in most cases be determined with the greatest accuracy when the beats are about four per second. When they are more rapid than this, they are difficult to count, and when they are slower, the exact moment of maximum or minimum intensity is less definite.

If we find by the method of beats the exact difference of frequency between each consecutive pair of forks over the whole octave, we can find the absolute frequency of any one of the forks. Let n be the frequency of the lowest fork, then that of the highest must be $2n$, as the interval is an octave. Next suppose that the number of beats between the first and second fork is a, that between the second and third is b, that between the third and fourth is c, and so on. Then the vibration numbers of the forks will be n, $n+a$, $n+a+b$, $n+a+b+c$ and so on, until we have for the highest fork $n+a+b+\ldots\ldots+z$. But we know that this is $2n$, and therefore n is the sum of all the beats throughout the series. Thus knowing n and all the terms a, b, c, etc., we can find the frequency of every fork of the series.

If now we wish to find the frequency of any other fork, which falls within the range of the tonometer, we have merely to compare it with the tonometer forks until we find one with which it beats, and by timing the beats find the difference between the frequencies of the two. This difference is then to be added to or subtracted from the known frequency of the tonometer fork to give that of the fork under investigation. In order to know whether we are to add or subtract the difference we must know which of the forks is the higher. When two forks A and B beat slowly, the ear cannot decide which is the higher in pitch. A simple method of finding which is the higher is as follows. Suppose the forks make four beats per second. Stick a very small piece of wax on the prong of B. We know that this will flatten the fork. If we find that the beats are now slower than four per second, we have made the difference of pitch less by flattening B, and therefore B was originally the sharper. If the beats are made quicker, B was the flatter of the two. Care must be taken not to stick too much wax on B. Suppose, for instance, we put on enough to flatten B by ten vibrations a second, then if B were originally four vibrations *above* A, we should flatten it to six vibrations *below* A. Thus we quicken the beats by adding wax to B, and if we apply our rule, we shall conclude erroneously that B was originally below A.

It is not necessary to flatten the fork by sticking wax on it when using the tonometer. It is always possible to find two consecutive forks of the tonometer with each of which the fork under investigation makes less than four beats per second, unless it happens to coincide exactly in pitch with one of the forks. It is clear that the fork under investigation must fall between these two tonometer forks in pitch. A determination of pitch by the tonometer is easily made, and takes little time, and the instrument is therefore much used by makers of scientific apparatus, bell founders, and musical instrument makers.

Appun constructed a tonometer in which the forks were replaced by Harmonium reeds. This form is more portable than Scheibler's, but it is less accurate, for the reeds influence each other to some extent. A reed has not quite the same pitch when sounded with the reed below it in the series, as it has when sounded with that above it.

234. Measurement of differences of frequency by Lissajous' Figures. The comparison of the frequencies of two notes of nearly the same pitch is most easily made by counting the beats in a given time when the two notes are sounded together.

A more accurate method depends on the use of Lissajous' Figures. The method is not conveniently applicable to organ pipes and the like, where air is the vibrating medium. It is most commonly used for comparing the pitches of tuning-forks, and it has been used in the investigation of the vibrations of a string, as will be seen in Chapter XVI. The arrangement of the apparatus is as shewn in Fig. 17, but it is better to observe the figures with a telescope instead of throwing them on a screen, as this allows a smaller source of light to be used, and the figures are thereby made sharper. The method has the advantage that it can be used, not only when the forks or other vibrating bodies are nearly in unison with each other, but also when they make very nearly one of the consonant intervals.

Suppose the method is applied to two forks nearly an octave apart; then the figure varies from a parabola with its vertex in one direction, through a figure of 8, to a parabola with its vertex in the opposite direction, and the time in which the higher fork makes one vibration more or less than double the number made by the lower is the time taken for the figure to change from one parabola to the other and back again. Let us suppose that the cycle of changes is found to be completed ten times in half a minute, then, if the frequency of the lower fork is 100, that of the upper must be either 199·67 or 200·33, for 10 cycles in 30 seconds means a gain or loss by the higher fork of one-third of a vibration per second. We can find whether we are to take the higher or lower number by the same method as was used with beating forks. Stick a very small piece of wax on the upper fork. This flattens the fork, and if it makes the changes in the Lissajous' figures slower, the fork is too sharp, and vice versa. Lissajous' figures are specially useful in tuning two forks accurately to some interval such as the fourth or fifth, for even though the interval is so nearly correct that the figure does not complete its cycle of changes before the vibrations die away, it is easy to see whether a change is taking place or not, when only a quarter or even less of the cycle can be observed whilst the vibrations continue.

235. Number of waves required to give the sensation of pitch. We have had ample evidence that we get a sensation of definite pitch only when a series of vibrations reach the ear at regular intervals, but there is one question which has not yet been answered. How many waves must reach the ear in order to define the pitch of a note? If, for instance, five waves with a period of ·01 sec. reach the ear, shall we hear a note of frequency 100; or do we need 50, or 500?

The question has been answered by Kohlrausch by means of a very simple experiment. He fixed below the bob of a pendulum a plate of metal in the shape of an arc of a circle with several teeth on its lower edge. Below the pendulum he placed a card in such a position that the teeth struck its edge in turn, as the pendulum swung across it. He then allowed the pendulum to swing once across the card, and varied the number of teeth to find the smallest number that would give a sensation of pitch.

He arrived at the surprising result that *two* teeth enabled him to form a fairly good estimate of the pitch of the note. From two to ten teeth the accuracy of the estimate increased, but above ten there was little, if any, further increase of accuracy.

236. Measurement of the velocity of sound in the open air. We shall turn next to the methods that have been used for measuring the velocity of sound in air and other media.

The earlier measurements of the velocity of sound in air were made by the direct method of observing the interval of time between the flash of a distant cannon and the arrival of the sound. This method suffers from several disadvantages. The velocity of sound depends on the temperature and humidity of the air, and on the wind; and all these may vary in some unknown way between the cannon and the observer.

The effect of wind can be eliminated to a great extent by the method of *reciprocal firing*. Two observers A and B are situated several miles apart, and each is provided with a cannon, and some means of measuring small intervals of time, such as a stopwatch or an electric chronograph. A fires his cannon, and B measures the time between the flash and the report. Then B fires his cannon, and A measures the interval between flash and report, and this operation is repeated a number of times at intervals of one or two minutes. The velocity calculated from A's observations will generally be found to be different from that calculated from B's observations. Suppose the wind is blowing from A to B, then the velocity found by B is the velocity of sound in still air plus the velocity of the wind, and that found by A is the velocity of sound minus the velocity of the wind. The mean of the two therefore gives the velocity of sound in still air.

There still remain the temperature and humidity of the air to be allowed for. These are observed at several points between the two observers, and correction made as explained in Chapter V

Another source of error is what is called the *Personal Equation* of the observer. It is impossible to record the time at the exact moment when the flash is seen or the report heard. The brain is a little late in realizing that the flash has occurred, and the hand is still later in making the record, if an electric recorder is used. A skilled observer is always late by about the same amount, so long as he is observing the same class of event. The actual magnitude of the personal equation is of no great consequence, so long as it is constant, and its amount is known, for in that case a correction can be made. Instruments have been constructed specially for the purpose of testing the constancy and finding the amount of the error made in recording the time of the occurrence of such events as the seeing of a flash, the hearing of a sound, the feeling of a touch or of an electric shock.

If the error were the same in recording the time of the flash of the cannon and the arrival of the sound, the interval between the two would be correct, and there would be no error in the velocity of sound calculated from the observations, so far as it is affected by personal equation. The error is, however, not generally the same for events of different classes. There would be in most cases a greater delay in recording the sound than in recording the flash.

Regnault tried to eliminate the error by making the flash and the arrival of the sound record themselves. A wire carrying an electric current was stretched across the muzzle of a gun. The explosion broke the wire, and the interruption of the current caused a mark to be made on a revolving drum. The arrival of the sound at the observing station was recorded by means of a stretched membrane, which was pressed forward by the arriving sound wave. The movement of the membrane broke a second electric circuit, and made another mark on the revolving drum. If the rate of revolution of the drum is known, the distance between the two marks gives the time taken by the sound in travelling from the gun to the observer. The effect of wind was eliminated by firing from each end alternately in the manner already described. Regnault's method did not get rid of personal equation entirely, for the membrane had a "personal" equation, as it did not respond immediately to the arrival of the sound-wave. The method had the advantage of substituting for a human delay a mechanical delay, which might be expected to be more regular. Regnault made special experiments for finding the amount of the error, and applied a correction for it.

As the results of all his experiments in the open air, it appears that the velocity of sound in still dry air at 0° C. is about 332 metres per second, but it is not independent of the intensity of the sound. A loud sound travels faster than a feeble sound. Regnault made experiments with the same gun over different distances, and it appeared that as the distance grew greater and the sound weaker, the velocity approached a limiting value of about 330 metres per second.

Bravais and Martin determined the velocity of sound at high altitudes in the Alps, and found the same value as at sea-level. Their experiments confirm the statement made in Chapter V that the velocity of sound is independent of the pressure, for the atmospheric pressure at the altitudes at which their experiments were made is much lower than at sea-level.

It may be mentioned as an instance of the great distance sound can travel before being extinguished that the Krakatoa eruption in 1883 sent out sound-waves which could be heard at distances of more than 2000 miles. The great wave caused by the main explosion, which destroyed the mountain, could be traced by the records of barometers for about five days, during which time it travelled more than three times round the earth.

237. Measurement of the velocity of sound in water. In 1823 Colladon and Sturm determined the velocity of sound in the water of Lake Geneva. The source of sound was a bell placed below the surface of the water. The lever that struck the bell ignited a charge of gunpowder, and the flash could be seen at the observing station. The moment of arrival of the sound was detected by means of a large ear trumpet, the small end of which was placed in the observer's ear. The large end was closed with a thin plate, and was placed under water. The velocity of sound in the water was found to be 1435 metres per second.

Threlfall and Adair found that the velocity of sound in water is very greatly affected by the intensity of the sound. The velocity of a loud sound may be as much as 35 per cent greater than that of a feeble sound.

238. Measurement of the velocity of sound in pipes. When sound travels through pipes, there is friction between the air and the pipe, and there is some slight transference of heat between the two. Both these causes tend to lower the velocity of the sound-waves, the reduction of velocity being greater the narrower the pipe, and the greater the wave-length of the sound.

Helmholtz and Kirchhoff investigated the matter theoretically, and both came to the conclusion that the difference between the velocity of sound of frequency N in free air and the velocity of the same sound in a pipe of diameter r is inversely proportional to r, and inversely proportional to the square root of N.

Regnault made an extensive series of observations of the velocity of sound in the newly laid water pipes in Paris in 1862, using such sources of sound as a pistol, an explosive mixture of gases, and various musical instruments. The method used was similar to that employed in his open-air experiments already described. He found as before that the velocity approaches a limit as the sound grows fainter, and the limit is lower for narrow pipes than for wide ones. The limiting velocity was the same for all the sources of sound. Some of the pipes had a diameter of 1·1 m. and Regnault concluded that for these the walls of the pipes had ceased to have any effect.

The limiting velocity of sound in these was therefore the same as in the open air, and was given by Regnault as 330·6 metres per second at 0° C.

239. Measurement of the velocity of sound by Resonance. Many indirect measurements of the velocity of sound in pipes have been made, use being made of the properties of organ-pipes described in Chapter X.

The simplest of the indirect methods depends on the measurement of the length of tube which gives maximum resonance with a fork of known frequency n. From the length of the tube we can find λ the wave-length of the note, and then, from the equation $v = n\lambda$, we find v.

The Resonance Tube is a plain cylindrical tube of glass or metal containing a closely fitting piston by which the length of the tube can be varied. Instead of using a piston we may immerse the tube to various depths in water; or the lower end can be connected with a vessel of water that can be raised or lowered, thus causing the water-level to rise or fall in the resonance tube.

Suppose the tube has been adjusted to the shortest length that gives resonance with the fork. The length of the tube plus the correction for the open end is now one quarter the wave-length of the note. Now lengthen the tube gradually. The resonance grows weaker until it almost disappears, and then increases again to a second maximum. The length of the tube plus the correction is now three quarters of the wave-length. Subtracting the first measured length from the second we have the half wave-length free from the correction for the open end, and so can calculate v from the formula $v = n\lambda$. The method has already been described in Chapter X as employed by Blaikley for finding the correction for the open end.

240. Measurement of the velocity of sound by Seebeck's Tube. Seebeck's Tube described in Chapter VII (Fig. 59) is in principle merely a modification of the Resonance Tube. When the side tube is at an antinode there is maximum motion but no variation of pressure at its inner end. Consequently no waves are sent to the ear, and no sound is heard. The distance from the centre of the side tube to the piston is then either a quarter of a wave-length, or a quarter of a wave-length plus one or more half wave-lengths. The procedure is similar to that with the simple resonance tube, with the exception that the points of silence are observed instead of the points of maximum sound and there is no correction for the open end to be considered.

241. Measurement of the velocity of sound by Organ Pipes.
Similar experiments have been made by Dulong, Blaikley, and others with organ pipes of adjustable length. The pipe is tuned to a fork of known pitch by observation of the beats. It is then lengthened until it gives the same note in the next higher mode of vibration. The difference between the two lengths is half a wave-length of the note of the fork.

242. Kundt's method of measuring the velocity of sound.
Kundt devised an accurate method of comparing the velocities of sound in different gases, or of comparing the velocities in the same gas in tubes of different diameters.

The later form of his apparatus is shewn diagrammatically in Fig. 96.

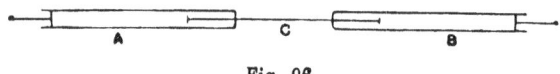

Fig. 96

A and B are two glass tubes, each of which has an adjustable piston at its outer end. The inner ends are closed by caps through which pass the ends of the glass rod or tube C. The rod is so adjusted that one quarter of its length is inside each of the tubes, and its ends are provided with circular discs a little smaller than the bore of the tubes.

As the rod is supported at the two nodes of its first overtone, it gives out this overtone, when rubbed lengthways near its middle with a piece of wet flannel. The discs on the ends of the rod communicate the vibrations to the air or other gas in the tubes A and B, and the air or gas is set into a state of stationary vibration. A little light powder, such as lycopodium powder or precipitated silica, is spread along the tubes. If the vibration is maintained for a short time, the powder leaves the vibrating segments where the motion is greatest, and gathers in small heaps at the nodes. We are thus able to measure the wave-lengths in the two tubes by measuring the distances between the heaps of powder—any two adjoining heaps being half a wave-length apart. The pitch of the note communicated to the contents of the tubes is the same for both, being the pitch of the note of the rod, and therefore the velocities of sound in the two tubes are proportional to the respective wave-lengths.

The pistons are provided to adjust the lengths of column of gas. If this is not done, the vibrations in the gas may be feeble and irregular, and the nodes ill defined. The adjustment is made by moving the piston back and forwards whilst the rod is rubbed, and so finding the position for which the powder is most strongly agitated.

A better way of forming the dust figures is to pour a small quantity of the powder through the tube, so that a narrow strip of dust is left along the bottom, the rest of the tube being clear. Now place the tube in such a position that the dusty strip is a little way up the side, and give a single stroke on the rod with the wet flannel. The dust will fall to the bottom of the tube everywhere except at the nodes, and at each node there will be left a sharply defined clear space, the position of whose centre can be estimated within a few tenths of a millimetre.

Kundt's apparatus is not well suited for finding the absolute velocity of sound in a gas, for in order to calculate this we need to know the frequency of the note. The glass rod gives a note of so high pitch that its frequency is not easily determined with accuracy. The apparatus is more commonly used for comparing the velocities in different gases. For this purpose one of the tubes contains air, and the other contains the gas to be compared with air. The note is necessarily exactly the same in the two gases, and if ordinary precautions are taken the temperature will be the same. Consequently the velocities in the two gases are in the same ratio as the measured wave-lengths. The velocity in air being known that in the other gas can be calculated.

The earlier form of Kundt's apparatus had only one tube, and the rod was held at its centre, so that it gave its fundamental note. Dry air was first introduced and the figures measured, and then the gas was introduced and the figures measured again. The objection to this form of the apparatus is that we cannot be certain that the note is of the same pitch in the two cases, for a small change of temperature alters the pitch of the rod appreciably, and further, it is not so easy to secure that the air and the gas to be investigated are at the same temperature, as it is with the double-ended form of the apparatus.

Kundt's apparatus has been used by many investigators for finding the ratio of the specific heats of a gas. The ratio is of importance in Thermodynamics, and gives valuable information about the constitution of the molecules of a gas.

We saw in Chapter V that the velocity of sound in a gas is $\sqrt{\dfrac{\gamma p}{\rho}}$, where ρ is the density of the gas at pressure p. Suppose the air and the gas investigated have the same pressure and temperature. We can then omit the factor p and we have

$$\dfrac{\gamma_1}{\rho_1} : \dfrac{\gamma_2}{\rho_2} = v_1^2 : v_2^2 = l_1^2 : l_2^2$$

or

$$\gamma_2 = \gamma_1 \dfrac{\rho_2}{\rho_1} \left(\dfrac{l_2}{l_1}\right)^2$$

where the suffix 2 refers to the gas, and the suffix 1 to air, and l_1, l_2 are the wave-lengths.

The gas need not be at the same pressure as the air when the experiment is made, for the velocity of sound in any one gas is the same at all pressures; but the values of the densities used in the formula must be those for corresponding states of the air and the gas. They may for instance be the densities at temperature 0° C. and pressure 760 mm., as given in the usual books of tables.

*243. Absolute measurement of the intensity of Sound. Some simple and accurate method of measuring the amplitude of the vibrations of sound-waves in air is much to be desired. Little has been done in this direction up to the present, and only a brief outline of the methods that have been used need be given.

*244. Rayleigh's method. Lord Rayleigh has formed an estimate in two ways of the amplitude of the faintest sound that is audible.

In the first experiment he used a whistle as his source of sound. The whistle was mounted on a bottle, and the pressure of the air in the bottle was kept constant. This pressure and the volume of air passing through the whistle per second were measured. The energy used per second in blowing the whistle could then be calculated, as it is equal to the volume of air passing per second multiplied by the pressure expressed in absolute units. It was found that the sound was just audible at a distance of 820 metres from the whistle. The bottle was on the ground in the centre of a large lawn; therefore at a distance of 820 metres the energy expended on the whistle was distributed over a hemisphere whose radius was 820 metres. Knowing the quantity of energy passing through each square centimetre of the hemisphere in the form of sound, it is possible to calculate the amplitude of the air vibrations. The amplitude thus calculated was found to be $8 \cdot 1 \times 10^{-8}$ cm. The actual amplitude must have been less than this, for it was assumed in the calculations that all the energy spent on the whistle was converted into sound—an assumption that is not quite correct as Lord Rayleigh pointed out, for some of the energy is spent in making eddies and in warming the air.

In a second experiment a tuning-fork and resonator were used as the source of sound, the rate of emission of energy for any amplitude of vibration of the fork being calculated from the rate at which the amplitude died down. With the exception of the method of calculating the rate of emission of energy the principle was the same as in the former experiment. The amplitude of the air-waves when the sound was just audible was in this case found to be $12 \cdot 7 \times 10^{-8}$ cm.

It is interesting to note as an illustration of the sensitiveness of the ear that $12 \cdot 7 \times 10^{-8}$ cm. is not much more than one five-hundredth part of the wave-length of green light.

Pellat has calculated that, if the quantity of heat given up by one gramme of water in cooling through $1°$ C. were entirely converted into electrical energy, and then were converted into sound by a good telephone, the energy would be sufficient to produce an audible sound for 10,000 years. Rayleigh finds that the stream of sound energy which is just able to affect the ear is of about the same order of magnitude as the stream of light energy which will just affect the eye.

*245. **Töpler and Boltzmann's Method.** Töpler and Boltzmann used optical interference to find the amount of condensation and rarefaction at a node in an organ pipe. Two opposite sides of the pipe were made of plates of glass, which projected above the closed upper end of the pipe. One of the two interfering beams of light passed through the plates of glass outside the pipe, and the other passed inside the pipe. The beams of light were brought together by a suitable optical method after passing the pipe, and formed a set of interference bands, which were observed with a telescope. If the air in the pipe is now compressed, its refractive index is raised, and the beam that passes through the pipe is retarded relatively to the other beam. Hence the interference bands are displaced, and from the amount of displacement the retardation of the beam and the increase in density of the air in the pipe can be calculated, if the relation between the refractive index and the density is known. When the pipe is sounding, condensations and rarefactions follow each other so rapidly that the consequent motion of the bands cannot be followed with the eye.

Töpler and Boltzmann used a stroboscopic method to make them visible. A tuning-fork of very nearly half the frequency of the pipe had a thin plate fixed to each prong, the plates being parallel to the plane of vibration and overlapping each other. In each plate was a slit parallel to the prongs of the fork, and so placed that the slits coincided when the fork was at rest. When the fork was vibrating, light could pass through the two slits only when the prongs were passing through their equilibrium positions, that is, twice in each vibration.

This fork was placed in such a position that the beam passing through the pipe could reach the telescope only by passing through the slits. If then both fork and pipe are vibrating, the beam reaches the telescope for a brief period twice in each vibration of the fork, or once in each vibration of the air in the pipe. If the pipe is exactly an octave above the fork, the vibration will always be in the same phase and the air will be in the same state of condensation, when the beam passes, and the interference bands will remain at rest with a displacement corresponding to that condensation. If however the ratio of the frequencies is not exactly 2 : 1, the relative phases will change slowly, and the bands will oscillate slowly from side to side. The maximum condensation or rarefaction can be calculated from the maximum displacement of the bands from their central position, and the law of variation of the density with the time from the motion of the bands.

Töpler and Boltzmann found that with a large pipe blown at a moderate pressure the variation from the mean density at the node amounted to ·009 of an atmosphere, and that the variation followed the harmonic law.

*246. Comparison of the intensities of two sounds.

The comparison of the intensities of two sounds of the same pitch is a simpler problem. Two of the methods that have been used for this purpose will be described.

Mayer placed a resonator before each of the two sources of sound to be compared, and joined the resonators by an india-rubber tube with a manometric capsule near its middle point. With this arrangement there is more or less interference of the sounds reaching the capsule by the two paths. If the sources of sound are at the same distance from the resonators, are equally intense, and are in exactly opposite phase, and if the two tubes leading from the resonators to the capsule are of the same length, there will be complete interference at the capsule, and the flame will be at rest. The sounds will not in general be in opposite phase, therefore we compensate this by altering the length of one of the rubber tubes until the jumping of the flame is a minimum. The interference cannot be made complete by this process alone, unless the two trains of waves have the same amplitude at the capsule, and this will be the case only when the sources have the same intensity. If the sources have not the same intensity, the stronger of the two is now moved away from its resonator until the flame comes to rest, when the intensities of the sources will be proportional to the squares of their distances from the resonators.

Lord Rayleigh has devised a method of comparing the intensities of sounds of the same pitch by making use of the hydrodynamical principle that a flat body held in a stream of fluid tends to set itself with its flat surface at right angles to the stream. An illustration of this principle is seen in the case where a light card is allowed to fall through the air. The card soon comes into a horizontal position. Here the card is moving and the air is at rest, but the forces tending to turn the card are evidently the same as when the card is at rest and the air is moving. The experiment can be varied by hanging the card by a string fastened to one corner, and holding it in a wind. The card will set its face to the wind.

The result is the same when the stream of fluid is not constant, but alternating; a surface oblique to the stream tends to set itself at right angles, whichever way the stream is flowing.

In an organ pipe we have an alternating current of air at the antinodes, and a light disc placed at an antinode will tend to set itself across the pipe.

Rayleigh used as a resonator a closed pipe whose first overtone had the same pitch as the two notes to be compared. When vibrations are set up in the pipe, there is a node at A, one third of the length from the open end, and an antinode at B, one third of the length from the closed end. At B there is suspended a light disc with a magnet on its back, and the pipe is put in such a position relatively to the earth's magnetic field, that the plane of the disc makes an angle of 45° with the axis of the tube. If now the air in the pipe is made to vibrate, the disc will turn, until the couple due to the alternating air currents is equal to the couple due to the displacement of the magnet from the meridian.

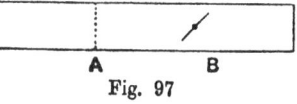

Fig. 97

The couple turning the disc from the meridian is proportional to the square of the mean velocity of the air, and therefore is proportional to the intensity of the sound for notes of the same pitch.

The two sounds to be compared are allowed to act separately on the resonator, and the respective deflexions of the disc are observed. The intensity of the vibration in the resonator is proportional to the intensity of the sound that causes it, and therefore the intensities of the two sounds are in the ratio of the tangents of the angle of deflexion.

CHAPTER XIII

THE PHONOGRAPH, MICROPHONE AND TELEPHONE

247. The Phonograph. The Phonograph was invented by Edison in 1877. It has been little altered from its original form, the general principle of the instrument remaining the same.

Fig. 98 shews the essential parts in a diagrammatic form. A is a conical mouthpiece by which the sounds to be reproduced are concentrated on a diaphragm B.

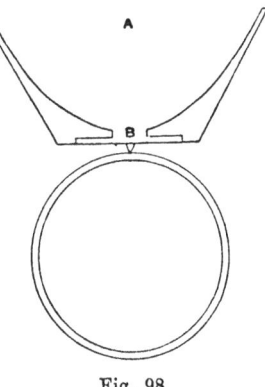

Fig. 98

The diaphragm, which may be made of glass, metal, mica, or other material, has a sharp style fixed to the centre of its lower surface, the pointed end of the style pressing against a rotating cylinder of hard wax. The cylinder is rotated at a uniform rate by clockwork or other means, and is mounted on an axis on which a screw is cut, so that it moves along its axis as it rotates. By this means the trace made on the wax by the style is in the form of a helix from end to end of the cylinder.

In some forms of the phonograph the trace is made on a flat circular plate which revolves round its centre, and the recording point is moved gradually outwards from the centre as the plate rotates, so as to cut a spiral curve in the wax.

The air-waves that enter the cone set the diaphragm in vibration, and the style cuts a furrow in the wax, the depth of the furrow varying from point to point in accordance with the varying displacement of the diaphragm.

When the cutting of the furrow has been completed, the cone and diaphragm are raised, and the cylinder returned to its starting point. The cone is then lowered again into its place, and the cylinder set in rotation. The style rises and falls as the depth of the groove varies, and makes the diaphragm repeat the vibrations it performed when the groove was being cut. The diaphragm communicates its vibrations to the air, and thus the features of the original sound are reproduced. In practice the cutting style is not used for reproducing the sound. It is so sharp that it will not rise and fall in the furrow, and soon destroys the trace. A style with a rounded point is therefore used to replace the cutting style, when the sound is to be reproduced.

The instrument copies the features of the original sound very closely. The voice of the speaker can be recognized, and the qualities of the notes of different musical instruments are easily distinguished. It fails to some extent when a sound is characterized by the presence of high harmonics, for the diaphragm cannot vibrate rapidly enough to record them on the cylinder.

The Phonograph and its modifications the Gramophone, the Pathéphone, etc. are becoming of great value, not only for reproducing music, but also for such purposes as teaching the pronunciation of foreign languages, and recording the speech and music of savage tribes. A phonograph now forms part of the outfit of every explorer.

The pronunciation of the languages of civilized nations is constantly changing, and it has been suggested that records should be made from time to time, and preserved for the use of future philologists. A museum of such records has already been established in Paris.

If speech is to be reproduced at the same pitch as the original, the cylinder must rotate at the same rate when the record is made, and when the speech is reproduced. If the cylinder is rotated more rapidly during reproduction, the pitch is raised. Not only does the pitch of the fundamental rise, but also the pitch of all the harmonics, and therefore the harmonic relations of the constituents is preserved, and the quality of the note is unchanged.

248. Test of vowel theories by the Phonograph. It would seem therefore that the phonograph provides a means of deciding between the two rival theories as to the nature of vowel-sounds.

Sing a vowel to a phonograph, and make a record in the usual way. Next reproduce the sound, but gradually increase the rate of rotation of the cylinder. All the constituents of the note given out will rise in pitch at the same rate, and the intervals between them will remain constant. The variable pitch theory requires this relation between the constituents of a vowel sound, when the pitch of the note on which the vowel is sung changes. If then this theory is true, the phonograph should give out the same vowel as was sung to it, at whatever rate the cylinder rotates. The fixed pitch theory on the other hand requires that some one or more constituents shall remain fixed in pitch, and therefore, if the theory is true, the phonograph should give out the original vowel only when the cylinder is rotated at the original rate.

Experiments made to test this point have proved to be quite inconclusive. Some observers say that the vowel changes when the rate of rotation changes, whilst others say it does not. The true physical cause of the differences between the vowels is still uncertain.

249. The Telephone. The Telephone has some features in common with the phonograph. The complete installation in its simplest form consists of two instruments, the transmitter and the receiver, connected by a pair of wires. In the transmitter the sound-waves cause vibrations in a flexible diaphragm as in the phonograph. These vibrations are converted into electrical waves, which travel to the receiver, and there produce vibrations in a second diaphragm, which in consequence gives out sound-waves similar to those entering the transmitter.

250. The Bell Telephone. The first practical telephone was invented by Graham Bell, and with modifications in details is still in use. The essential parts are shewn in Fig. 99.

A is a permanent magnet with one of its poles placed close to a thin iron diaphragm B. Round this pole are wound many turns of fine wire, and the two ends of the coil are joined to another similar instrument.

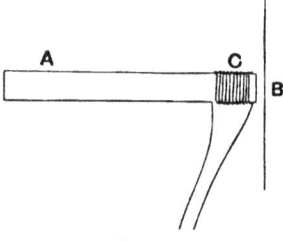

Fig. 99

Let us suppose the right-hand end of the magnet is the north pole, and the diaphragm has been removed. We have now lines of magnetic force starting from the N. end, passing through the air to the S. end, and then returning through the material of the magnet to the N. end.

The coil C is placed close to the end of the magnet, and most but not all of the lines of force pass through it. Some lines escape from the magnet before they reach the coil, and others come out through the sides of the coil. If we now bring up the diaphragm, we shall alter the distribution of the lines of force. The metal of the diaphragm forms an easier path for the lines than does the air, and some lines which escaped from the sides of the magnet will now remain inside until they reach the end, will there cross the air gap and run along the diaphragm for part of their paths.

The result is that the coil now encloses more lines of force than before, and the closer the diaphragm is to the pole of the magnet, the greater will be the number of lines passing through the coil. Hence, if the diaphragm is set vibrating by sound-waves, the number of lines passing through the coil will rise and fall in accordance with the vibrations.

When the number of lines of force through a circuit is changing, an electromotive force is set up, and this produces a current in the circuit. The strength of the current depends on the rate of change in the number of lines through the circuit, and the direction of the current is different according as the number is increasing or decreasing. Consequently the vibrations of the diaphragm will set up alternating currents in the coil, and the magnitudes and directions of the currents will depend on the nature of the vibrations.

We have now to consider what happens when these currents produced by the voice in the transmitting instrument reach a similar receiving instrument.

If a coil of wire is wound round a bar of iron and a currrent is sent through the coil, the iron becomes magnetized, and remains so as long as the current is flowing. The intensity of the magnetization increases with the strength of the current, and the direction of the magnetization is reversed if the current is reversed. If the iron bar is a permanent magnet, and the current is feeble, the magnetizing effect of the current may be insufficient to reverse the magnetization. In this case a current in one direction will increase the intensity of magnetization and a current in the other direction will diminish it. This is what happens in the receiving instrument. The alternating currents produced by the transmitter cause corresponding changes in the strength of the magnet of the receiver. The diaphragm of the receiver, being made of iron, is drawn inwards by the magnet, and the extent to which it is drawn inwards depends on the strength of the magnet; and this again on the strength and direction of the currents in the circuit. Thus we see that the vibrations of the receiver diaphragm will be controlled by those of the transmitter diaphragm, and the air vibrations near the transmitter will be reproduced in the air near the receiver.

The Bell telephone then will serve both as transmitter and receiver. It has been superseded by other forms of transmitter, but with some alterations in form is in practically universal use as a receiver. The principal change is the substitution of a horse-shoe magnet for the bar magnet. In the receiver most commonly used in this country, the horse-shoe magnet has its two prongs parallel and close together. Each prong has a short soft iron armature attached, and a flat coil of many turns of fine wire is wound round each armature. The outer ends of the two armatures are side by side and close to the diaphragm. This telephone is the same in principle as the form we have described, but it is more effective in producing vibrations in its diaphragm, for, as the two poles of the magnet are near each other, the magnetic field in their neighbourhood is much stronger than when they are at opposite ends of a bar.

A pair of such telephones can be used as transmitter and receiver only for short distances. The energy of the vibrations of the receiver disc is many thousand times less than that of the transmitter disc, and if many miles of wire are included in the circuit, the currents are so much reduced by the added resistance, that the sounds become inaudible. Consequently a different principle has been adopted for the transmitter, in order that stronger currents can be used.

251. The Hughes Microphone. Almost all the transmitters in use at the present time are modifications of the Hughes microphone. This instrument is shewn in section in Fig. 100.

A is a thin board, on which are fixed two blocks of carbon B and C.

Each of these blocks has a conical hole bored in one side. Resting loosely with its pointed ends in the two holes is a rod of carbon D. Two binding screws on the back of the board are connected each with one of the carbon blocks, and a battery and telephone in series are joined to the terminals. The current then passes two points where the rod rests loosely against a carbon block. The resistance of a point of contact between two carbon surfaces changes greatly, when the pressure between the two surfaces changes, and therefore, if the microphone is shaken, and the pressure of the rod on its supports thereby altered, changes of intensity will be produced in the electric current, and the telephone will give out sounds.

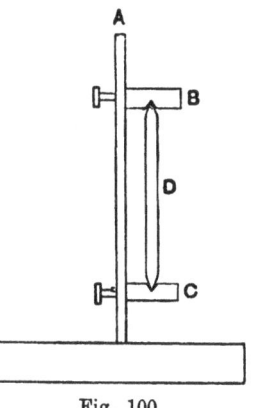

Fig. 100

The microphone is extremely sensitive to slight movements.

If a feather is drawn lightly across the board A, a loud rasping sound is heard in the telephone, and the footfall of a fly walking on the instrument is quite audible.

252. The Carbon Transmitter. Such an instrument, exactly as it has been described, could be used as a transmitter. If one speaks close to the microphone, the words can be heard in the telephone, but only faintly, and it has been modified in several respects to increase its efficiency.

The first change made was to send the current from the microphone through the primary of an induction coil, and to use the current from the secondary to work the distant telephone. The secondary current has a much higher electromotive force than the primary, and is less weakened by being transmitted through great lengths of line wire.

The next improvement was to increase the number of points of contact of the carbons, so as to reduce the resistance, and at the same time to intensify the changes of resistance. In one type of transmitter this is done by using several carbon rods arranged in parallel so that the current is divided amongst them. The rods are attached to a thin piece of wood, which forms the side or top of a box, and the sound-waves cause vibrations in the wood and the attached carbon rods.

The Gower, Crossley, and Ader Transmitters are instruments of this type. Such instruments have been extensively used in the past, and are still used for such purposes as transmitting music from a concert room, but for ordinary purposes they are now almost entirely superseded by transmitters which make use of granular carbon for the variable contacts.

253. The Granular Transmitter. The instrument shewn in Fig. 101 illustrates the principle of the largely used granular transmitters.

A wooden case about 3 inches in diameter has in it a recess A. At the bottom of the recess is a thin carbon plate B connected with a binding screw

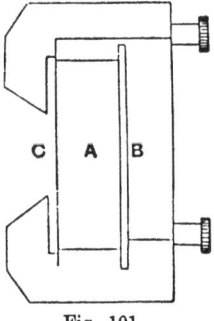

Fig. 101

at the back. In front of the recess is a diaphragm C made of platinum foil, and connected with the second binding screw at the back of the instrument.

The space between the platinum and carbon plates is nearly filled with carbon ground to a rather coarse powder.

The primary current of the induction coil passes through the granular carbon from the carbon plate to the platinum foil, and as there are a large number of points of contact between the grains, vibrations of the platinum plate caused by sound-waves give rise to large variations in the intensity of the current. The secondary current from the induction coil is carried by wires to the receiving instrument.

CHAPTER XIV

CONSONANCE

254. The Harmonic Series. In Chapter I we spoke of certain intervals, the octave, fifth, fourth, etc., as being worthy of special names, the chief reason for this preference being that they are pleasing to the ear and are easily recognized. In Chapter II these intervals were grouped together into the Harmonic Series. At a later stage we found that the proper tones of stretched strings and of narrow organ pipes form the same series. Finally we saw by Ohm's Law that a simple harmonic vibration is the proper unit to take in analysing complex notes, and by Fourier's Theorem that any complex note can be resolved into simple harmonic constituents that are members of the Harmonic Series. Thus it appears that we have weighty reasons for regarding the series as of fundamental importance.

255. Consonant Intervals. The intervals between the lower members of the series have vibration ratios which can be expressed by the ratios of small integers, and it is found that the smaller the integers which express the vibration ratio, the smoother and more pleasing is the corresponding interval to the ear.

The octave is given by the vibration ratio $2:1$, which involves smaller whole numbers than any other ratio except the unison $1:1$. No other interval gives an effect approaching that of the octave in smoothness. It is in fact so smooth and free from character, that it is hardly regarded in music as a harmony, but rather as doubling or intensifying the effect of either of the two notes which form the octave. The feeling of sameness is so great in the case of octaves, that though it is easy to decide whether two notes are or are not an exact number of octaves apart, it is often difficult to decide how many octaves there are between them.

CONSONANCE

The next interval in the series is the fifth 3 : 2. This again is a very good concord. It is smooth and pleasant to the ear, but lacks the sensation of sameness that is felt with the octave.

The sensation of smoothness gradually diminishes as we rise in the series through the fourth, major third, minor third, etc., until, when we reach the interval 9 : 8, the effect becomes unpleasant.

Helmholtz was the first to give a physical explanation of the various degrees of consonance of the different intervals and the remainder of this chapter will be devoted to giving an outline of his theory.

256. Effect of the beating of two pure tones. We must first consider the varying effect produced by the beating of two pure tones, when the interval between them is gradually widened, as Helmholtz's theory is to a great extent based on this effect.

Take two tuning-forks of the same pitch mounted on resonance boxes, and lower the pitch of one of them by sticking a little wax on its prongs, so that when the two forks are sounded together, beats are produced. If there are not more than three or four beats a second the effect is not unpleasant. A similar tremulous sound is produced intentionally by singers, and is imitated in certain stops of the organ and harmonium. If the lower fork is weighted still more, the beats become more rapid, and begin to be disagreeable, and the unpleasantness increases with the rapidity of the beats up to a certain maximum. As the rapidity increases beyond that corresponding to the maximum of roughness, the beats become less easily distinguishable, and the disagreeable effect diminishes, until, when the beats are so rapid that the ear no longer hears the alternations in intensity, no roughness is left.

We have an analogous case in the effect of a flickering light on the eye. If the rise and fall of intensity is slow, it is not unpleasant, and if it is very rapid, it is not detected at all. There is an intermediate rate that is very unpleasant and fatiguing to the eye. An effect of this kind is often experienced when watching an inferior Kinematograph exhibition. If the successive pictures do not follow each other quickly enough on the screen, the flicker soon fatigues the eye.

It will not be possible by merely sticking wax on the prongs of the fork to get a sufficient range of alteration of pitch to shew the whole variation of roughness from unison through the maximum of roughness to the stage at which the beats cease to be perceptible. A greater range can be secured by using a tuning-fork with sliding weights which can be clamped at any part of the prongs. The nearer the weights are to the free ends, the lower is the pitch. A still better method of shewing the effect of the varying rapidity of the beats is by blowing together two similar open organ pipes of wide bore, and lowering the pitch of one of them by gradually covering the open end. A wide range of tuning is possible in this way, and the interval between the pipes can be made so great that the roughness disappears.

257. Rate of beating which gives the worst dissonance. Helmholtz and others have made experiments to find at what rapidity of the beats the roughness reaches a maximum, and at what rapidity it disappears again.

The result is not the same at all parts of the musical scale. Near the middle of the pianoforte scale the dissonance reaches a maximum at about 32 beats per second, and disappears at about 85. These numbers of beats correspond to intervals of about a semitone and a minor third respectively.

Higher in the scale the number of beats that gives maximum dissonance increases, though not in proportion to the frequency, and the interval that gives the worst effect diminishes. Similarly for low notes the worst number of beats is less than 32, and the worst interval is greater than a semitone. Sound together on the pianoforte or harmonium two notes a semitone apart. If they are near the bottom of the scale the effect is not very unpleasant—not nearly so bad, for instance, as a major third in such a position. In the middle of the scale a semitone is exceedingly unpleasant. Near the top of the scale it is still unpleasant, but not so unpleasant as near the middle. At the top of the scale the worst interval is less than a semitone. If the instrument used made it possible to sound together here two notes a quarter of a tone apart, their effect would be much worse.

Over several octaves near the middle of the scale the most

dissonant interval is not far from a semitone, and we shall assume it to be a semitone in all that follows; remembering however that there is some roughness up to an interval of a tone or more.

258. Beats due to difference tones. When the interval between two pure tones sounded together is increased beyond the interval at which the roughness disappears, the beats formed in the way we have just described continue to get more rapid, and the jarring effect does not reappear. Faint beats are heard when the two notes are nearly a fifth or nearly an octave apart, but these are caused by the combination tones. Let us suppose the frequencies of the two notes are 400 and 795, the interval being thus a little less than an octave. The first Difference Tone has a frequency $795 - 400$ or 395 and therefore makes 5 beats a second with the lower primary. If two forks making an interval of a mistuned octave are placed close together and made to give out loud notes, the beats can be heard plainly.

The beats of a mistuned fifth are much fainter, as they are dependent on a second Difference Tone. Suppose the two forks have frequencies 400 and 598, so that the higher is 2 vibrations a second too low to form a perfect fifth. The first Difference Tone has a frequency $598 - 400 = 198$. There are two second Difference Tones of frequencies $598 - 198$ and $400 - 198$ respectively. The former coincides with the lower primary; the latter beats 4 times a second with the first Difference Tone. Second Difference Tones are always faint and therefore the beats are very feeble.

259. Helmholtz's Theory of Consonance. Helmholtz's Theory of the physical basis of consonance and dissonance ascribes all dissonance to the roughness caused by beats. According to this theory the chord formed by two or more notes sounded together is the more consonant, the more free it is from beats of such a rapidity as causes roughness.

Applying the theory to the case already discussed—that of chords formed by two pure tones—we may say that from unison to a major third there is a varying amount of dissonance, the maximum dissonance being at the interval of a semitone for chords near the middle of the pianoforte scale. As the interval between the two notes is increased beyond a major third, the dissonance does not reappear, except very faintly near the fifth, and somewhat more perceptibly near the octave. Pure tones however are not much used in music. It was mentioned that certain stops in the organ give nearly pure tones. In the orchestra the only instrument that gives notes approximating in the least to pure tones is the flute, and even in the notes of the flute the octave is quite perceptible.

When a chord is formed from two or more complex notes we have to consider not only the beats of the fundamentals but also the beats arising from approximation of the harmonics of the various notes to each other.

When harmonics are present, a single complex note may have in itself the elements of dissonance. The harmonics are closer together, the higher we ascend in the series, and in the neighbourhood of the tenth or eleventh they come within beating distance of each other. If then a note has a long range of powerful harmonics, it will be more or less harsh. It is this prominence of the higher harmonics that gives a rough and penetrating quality to the notes of the trombone and hautboy.

In most cases the harmonics fall off so rapidly in intensity as we rise in the series, that it is usual in treating of the causes of dissonance to ignore all above the sixth, as being too weak to have any appreciable effect.

In the course of the discussion we shall have to take account of the fact that whilst some instruments, such as strings and open pipes, have the full series of harmonics, others, such as stopped pipes, have only the odd members. When two notes are sounded together on two instruments, one from each class, it is not always a matter of indifference which instrument plays the upper note.

260. Analysis of the consonant intervals. We shall first discuss the relative consonances of the more usual musical intervals produced by two instruments, each of which has the full series of harmonics, and shall ignore all harmonics above the sixth.

It sometimes happens that a composer wishes to produce

a very harsh effect, and therefore writes a chord containing two notes whose fundamentals are within the beating distance of each other. Suppose for instance he directs that the two notes B and c, a semitone apart, are to be sounded together. The two series of harmonics are as in Fig. 102 and it is seen that each harmonic of B is a semitone from the corresponding harmonic of c, and therefore the roughness is made greater by the presence of the harmonics.

Fig. 102

Fig. 103

Fig. 103 gives the scheme of harmonics for the Octave. In this case the upper note introduces no harmonics that were not already present in the lower, and so no roughness is produced. The consonance of the octave is perfect.

Next take the Fifth (Fig. 104). Here the consonance is not quite so good, for the 3rd harmonic of the higher note is within a tone of the 4th and 5th harmonics of the lower. This does not introduce much roughness, for the 4th and 5th harmonics are not generally very strong; and a tone is greater than the interval which gives maximum roughness. Hence the Fifth, though not perfect, is a good concord.

Fig. 104

Fig. 105

The Fourth (Fig. 105) is distinctly worse than the Fifth. The clashing of two harmonics a tone apart is lower in the series than was the case with a Fifth, and is therefore more conspicuous and we have also the 5th harmonic of the lower note making the bad interval of a semitone with the 4th of the upper.

The Major Third (Fig. 106) is still worse. We have now two intervals of a semitone included in our range of 6 harmonics.

The Minor Third (Fig. 107) is much the same as the Major Third. The 4th harmonic of the lower note is a tone distant from the 3rd of the upper, whereas the interval in the corresponding position was a semitone in the case of the Major Third, but on the other hand the semitone between E and $E\flat$ is a step lower in the series. We shall see presently that the difference between the characters of the Major and Minor Thirds is largely dependent on the existence of Differential Tones.

Fig. 106   Fig. 107

It is not necessary to examine in this way all the consonant intervals used in music, as the student should find no difficulty in working out the Major and Minor Sixths, the Minor Seventh, etc., for himself, but a few cases of special interest will be added to those given above.

261. Effect of widening an interval by an octave.
The consonance of an interval does not generally remain the same, when that interval is increased by an octave. Let us compare, for instance, a Fifth with a Twelfth.

Fig. 108 gives the harmonics for a Twelfth, and it will be seen that the consonance is perfect, as there is no clashing whatever, and this is true however high we go in the series. A Fifth then is made a perfect consonance by being widened by an octave. It is clear that the consonance of any interval formed by the lowest and any other member of the harmonic series must also be perfect.

Fig. 108

Fig. 109

Take next the Minor Tenth for comparison with the Minor Third. Referring to Fig. 107 we see that the dissonance has been somewhat increased by widening the interval. In the case of the Minor Tenth the 2nd harmonic of the higher note is within a semitone of the 5th of the lower note, whereas in the case of the Minor Third it was the 4th of the higher which formed a semitone with the 5th of the lower. The 2nd harmonic is generally stronger than the 4th, and the increased roughness due to this has more effect than the gain in smoothness arising from the absence from the Tenth of the tone interval found in the Third.

It will be found that most intervals change in consonance when extended by an octave, some becoming more smooth and others less smooth. Helmholtz makes the general statement that, if the smaller number is even, when the vibration ratio of an interval is expressed by the smallest possible integers, the consonance will be improved by widening the interval by an octave, and if the smaller number is odd, the consonance will be made worse. The Fifth whose ratio is 2 : 3 belongs to the first class, and the Minor Third whose ratio is 5 : 6 belongs to the second. Helmholtz's rule is merely an extension of the general statement made earlier in this chapter that the smaller the integers expressing the vibration ratio, the smoother is the interval. Suppose we widen the interval by putting the lower note down an octave. If its number was originally even, no fractions are introduced by halving it, and one of the integers expressing the new vibration ratio is smaller than before whilst the other is unchanged. If the lower number is odd, both numbers expressing the vibration ratio must be doubled before we can halve the lower without introducing a fraction. Thus in this case the lower integer remains unchanged and the higher is doubled.

262. Consonance of an interval formed by two notes which have not both the full series of harmonics. All the conclusions come to in §§ 260 and 261 rest on the assumptions (1) that each of the notes has the full series of harmonics, and (2) that the harmonics above the sixth are so weak as to be negligible. If these conditions are not satisfied, the results may be different. If for instance the notes are sounded on two stopped pipes, which have only the odd harmonics, the consonance is generally better than when the full series is present. The student can easily verify this for himself.

If one of the notes has the full series, and the other has only the odd members, there may be a difference in smoothness according to which of the notes is the higher. We will give one instance of such a difference. The hautboy has a conical tube and gives the full series of harmonics, whilst the clarinet has a cylindrical tube, and gives only the odd members of the series. Suppose the two instruments play two notes making an interval of a Major Third, first with the hautboy above the clarinet and secondly with the positions reversed.

Fig. 110 shews the two cases, and it is clear that there is a great difference in consonance. When the hautboy takes the higher note there is no clashing within the first six harmonics, whilst when the clarinet is above the hautboy, there are two pairs of notes making the interval of a semitone.

The student may test his knowledge of the method by finding the best position for the

Fig. 110

instruments when the interval is a fourth. He will find that in this case the clarinet should play the upper note.

The effect is not so striking in practice as might appear from what has been said, for in the case of the clarinet, and still more of the hautboy, the higher harmonics are so strong, that those above the sixth cannot be ignored.

263. Effect of combination tones formed by the harmonics. The harmonics of two notes will, if powerful enough, generate Combination Tones with each other, and it might be thought that this would introduce new tones, and so modify the conclusions. The harmonics are generally too weak to generate any but the first Difference Tones, and even these are not strong enough to have much practical effect on the consonance. Helmholtz has shewn, moreover, that, when the full series of harmonics is present, the first Difference Tones cannot generate beats, except when beats of the same frequency are already present from the clashing of the harmonics themselves, and therefore there can only be a slight increase in the strength of the beats. If however both notes contain only the odd harmonics, the Difference Tones may introduce the even terms, and so have some slight effect on the consonance.

We are for the present not taking account of the Difference Tones produced by the fundamentals. We shall see later that they have some effect in modifying the character of a consonance, though they may not cause beats.

264. Consonant Triads. The intervals less than an octave which are admitted as consonant in music are the following :—

Interval	Vibration Ratio
Fifth	3 : 2
Fourth	4 : 3
Major Third	5 : 4
Minor Third	6 : 5
Major Sixth	5 : 3
Minor ,,	8 : 5

Let us see how we can add these intervals in pairs, so as to give chords of three notes, each of which forms a consonant

234   CONSONANCE   [CH. XIV

interval with both the others, and such that the interval between the highest and the lowest notes of the chord is less than an octave. Such a chord is called a *Consonant Triad*.

Take, for instance, two notes which we may call p and q, making one of the consonant intervals with each other, and a third note r, making one of the consonant intervals with q, p being the lowest and r the highest in pitch of the three. We have then a chord of three notes p, q and r, where p to q and q to r are consonant intervals. If it prove that p to r is also consonant, and less than an octave, the three notes form a Consonant Triad.

We can make the following table giving all the pairs of consonant intervals whose sums are less than an octave:

1 Minor Sixth + Minor Third gives $\frac{8}{5} \times \frac{6}{5} = \frac{48}{25}$

2 Fifth   + Major Third ,, $\frac{3}{2} \times \frac{5}{4} = \frac{15}{8}$

3 Fifth   + Minor   ,,   ,, $\frac{3}{2} \times \frac{6}{5} = \frac{9}{5}$

4 Fourth   + Fourth   ,, $\frac{4}{3} \times \frac{4}{3} = \frac{16}{9}$

5 ,,   + Major Third ,, $\frac{4}{3} \times \frac{5}{4} = \frac{5}{3}$

6 ,,   + Minor   ,,   ,, $\frac{4}{3} \times \frac{6}{5} = \frac{8}{5}$

7 Major Third + Major Third ,, $\frac{5}{4} \times \frac{5}{4} = \frac{25}{16}$

8 ,,   ,, + Minor   ,,   ,, $\frac{5}{4} \times \frac{6}{5} = \frac{3}{2}$

9 Minor ,,   +   ,,   ,,   ,, $\frac{6}{5} \times \frac{6}{5} = \frac{36}{25}$

We find that only three of the resulting intervals are consonant, the 5th, 6th and 8th, which are a Major Sixth, a Minor Sixth and a Fifth respectively. Each of these combinations gives two Consonant Triads, for the Major Sixth, for instance, can be made up of a Major Third above a Fourth, or of a Fourth above a Major Third. We have then the following chords:

Major Third	above	Fourth
Fourth	,,	Major Third
Minor Third	,,	Fourth
Fourth	,,	Minor Third
Minor Third	,,	Major ,,
Major ,,	,,	Minor ,,

265. Derivation of the Consonant Triads from two fundamental forms. In musical notation the triads can be written as follows:

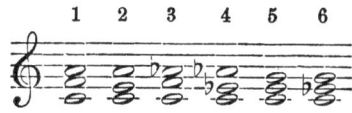

Fig. 111

These are the only groups of three notes within the compass of an octave that form consonant intervals with each other. They can of course be transposed into any other key. We are concerned here only with the relations which the notes in any triad bear to each other.

The first four of the triads can be derived from the 5th and 6th in a simple way. Raise the lowest note of 5 an octave and we get [chord] which is the same as 4, since it is a Fourth above a Minor Third. Now raise the lowest note of this triad an octave, and we get [chord] which is the same as 1. Similarly raising the lowest note of 6 an octave gives 2, and again raising the lowest note we get 3.

Hence we may write the Triads in two groups in accordance with this method of derivation.

The two groups are termed Major and Minor Triads respectively, because in the first or fundamental form of the

236   CONSONANCE   [CH. XIV

first group we have a Major Third between the two lowest notes, whilst in the second group we have a Minor Third.

Fig. 112

In each group the first chord is called the Common Chord —Major or Minor as the case may be—of the lowest note of the chord. The second chord in each group is called the First Inversion of the Common Chord, and the third is called the Second Inversion.

266. Consonance of the triads. The consonances of these chords can be investigated in the same way as the chords of two notes. It is found that the Second Inversion of the Major Triad is the smoothest, and the Second Inversion of the Minor Triad is the roughest of the six chords. The scheme for these is given in Fig. 113.

Fig. 113

The difference between the two chords is very marked. There are only two semitone intervals in the first group, and these are fairly high in the series, whilst the second group contains five. The student should be able to work out the other four triads in the same way without difficulty.

We can estimate the consonance of the triads still more simply by considering what chords of two notes enter into them, and making use of what has been found earlier with regard to the consonance of chords of two notes.

The first Triad of Fig. 111 contains a Fourth, a Major

Third and a Major Sixth; the third contains a Fourth, a Minor Third, and a Minor Sixth. The Fourth is common to the two, the Major Third is not greatly different from the Minor Third in smoothness, but the Minor Sixth is decidedly worse than the Major Sixth, and therefore the third group is worse than the first.

If we estimate the consonance of the fundamental forms of the Major and Minor Triad in this way, we find no difference between them, for each contains a Major Third, a Minor Third and a Fifth; yet there is no doubt that the Minor Triad is less harmonious than the Major Triad. Helmholtz ascribes the difference in harmoniousness to the Difference Tones formed by the three fundamentals of each triad taken in pairs.

Fig. 114 shews the positions of the First Difference Tones

Fig. 114

for each of the six chords. The chords are shewn in Minims, and the difference tones in Crotchets. A difference tone written with two tails can be obtained in two ways. Thus, for instance, the relative frequencies of the three notes in the first chord are $4:5:6$. The difference tone of the two lowest notes has a frequency $5-4$, and that of the two upper notes has a frequency $6-5$. The two Difference Tones therefore coincide at a frequency 1, which is two octaves below the lowest note of the triad. The highest and lowest notes of this first chord give a difference tone of frequency $6-4$, which is one octave below the lowest note of the chord. The difference tones of the remaining five chords are easily obtained in a similar way.

The difference tones of the Major Triads introduce no notes extraneous to the chords. They merely double in a different octave notes already present. The difference tones of the Minor Triads are, it is true, not within beating distance of each other or of the notes of the chords, but some of them fall quite outside the harmony. They are not strong enough to give the character of dissonance, but they disturb the harmoniousness of the chords, and give the Minor Triad its peculiar veiled and mysterious effect.

In former times it was not uncommon to use a major chord as the concluding chord of a composition, that was elsewhere in the minor key. This is not often done now. It was no doubt due to musicians of past times regarding the Minor Triad as hardly deserving the name of a consonance, and as not being sufficiently satisfying to the ear to be worthy to take its place as the final chord.

It should be reiterated that these conclusions are true only of just intonation. When, as is more usual, tempered intonation is used, the difference between the major and minor chords is to a great extent obscured by other causes due to the mistuning of the intervals.

CHAPTER XV

DEFINITION OF INTERVALS. SCALES. TEMPERAMENT

267. Definition of intervals. We must now return to chords of two notes, and find to what extent an interval may be mistuned without introducing unpleasant elements into the concord. In all systems of Temperament some of the intervals are a little out of tune, and it is important, when devising such a system, to know which intervals must be correct or nearly so, if dissonance is to be avoided, and which intervals can be modified without serious ill-effect. We shall find that in most cases beats are produced when an interval is a little out of tune, and the stronger these beats are, the more accurately must the interval be tuned. An interval for which the beats due to mistuning are strong is said to be sharply defined.

268. Definition of intervals formed by pure tones. The beats of mistuning generally arise from the harmonics, but in some cases they are caused by combination tones. In the case of pure tones there are no harmonics above the first, and therefore such definition as exists must arise from the combination tones. We have already seen, when discussing the consonance of intervals formed by pure tones, that, as the interval is gradually increased, we have first powerful beats due to imperfect unison. These beats get more rapid, until at an interval of about a Minor Third they cease to be perceptible as beats. They reappear when we are getting near a Fifth, and get gradually slower, until at the interval of an exact Fifth, they disappear. A Fifth between two pure tones then is defined to some extent, since beats are introduced by mistuning, but as the beats arise from the clashing of a first and a second difference tone, they are very faint, and therefore the definition of the interval is slight.

240   DEFINITION OF INTERVALS   [CH. XV

Near an Octave the beats are stronger, for they arise from the proximity of the first difference tone to the lower of the two primaries. The Octave then is fairly well defined, even when the notes are pure tones. It is easy to adjust the interval between two tuning-forks on resonance boxes to an exact octave by raising or lowering the pitch of one of them until the beats disappear. It is difficult to tune a Fifth in this way, as the beats are too faint.

The remaining consonant intervals of pure tones within an octave give no perceptible beats when mistuned, and therefore cannot be said to be defined at all.

269. Definition of intervals formed by complex notes. We turn next to the intervals formed by notes having the full series of harmonics, and limit ourselves as before to the first six harmonics.

It is seen in Fig. 103 that, when the two notes make a true Octave, every harmonic of the upper note coincides with a harmonic of the lower. In particular, the lowest harmonic or fundamental of the upper note coincides with the 2nd harmonic of the lower note. These are generally very strong, and therefore, if one of the notes is put a little out of tune, powerful beats will be heard.

The Octave is in fact so sharply defined that it is not possible to put it out of tune to the slightest extent without causing conspicuous beats. Consequently, in any system of tuning all Octaves must be true. The Fifth (Fig. 104) is also well defined, though not so well as the Octave. The 2nd harmonic of the higher note coincides with the 3rd of the lower, and these beat if the interval is not exactly in tune. The beats are easily heard, and the interval can be accurately tuned by making use of them. They are not however strong enough to be very obtrusive, and if the mistuning is slight, so that they are slow, they do not have any serious effect on the consonance. In the modern system of tuning keyed instruments all the Fifths are very slightly flatter than the true Fifth.

The Fourth (Fig. 105) is still less sharply defined than the Fifth, as the harmonics which coincide are now the 3rd of one series and the 4th of the other.

In the case of the Major and Minor Thirds and Sixths the intervals are very weakly defined, as the lowest coinciding harmonics are too high in the series to be of much consequence. These intervals can be mistuned to a considerable extent without suffering much in consonance, and on modern instruments they do in fact differ appreciably from the true intervals.

We see then that not only does the quality of a note depend on the presence of its harmonics, but also that the precision with which intervals must be tuned to avoid beats depends on these harmonics.

270. Tonic relationship. Modern music depends largely for its effect on the relation of the various notes to some one note which is called the *Tonic*. The tonic is what one might term the centre of gravity of the music. After a few notes of a melody have been sung, or still more noticeably after a few chords of harmony have been played, one feels that the music centres round some tonic. The melody usually comes to an end on the tonic, and the last chord is almost invariably a major or minor chord with the tonic as its lowest note. This is so very commonly the case, that one is hardly ever led astray by taking the last note of the bass as the tonic of the piece. The close is more marked and feels more restful when both the melody and the bass part end on the tonic. If the composer adopts some other ending, it may generally be taken to be due to his wishing to secure a less restful finish. A melody sometimes ends on the third or fifth above the tonic, and the peculiar effect of such an ending accentuates the feeling of tonic relationship. Since then the tonic relationship occupies such a prominent place in modern music, it may be anticipated that anything which weakens one's appreciation of intervals will detract from the effect of the music. This is the case when pure tones are used. The tonic is identified by the intervals between it and the other notes of the scale, and the appreciation of these intervals is less precise when the higher harmonics are weak or absent.

There is an instrument called the Ocarina which produces notes exceptionally free from harmonics. Anyone who, like the writer, has had the opportunity of hearing a quartet of four Ocarinas, will understand why such instruments are not used in the orchestra. The general effect is soft and smooth and for a short time is pleasant, but it very soon becomes monotonous. The harmony is quite colourless, and does not seem to be made much worse when, as is often the case, the notes are out of tune. A similar effect is noticed when a hymn tune is played on the organ on some stop such as a Stopped Diapason, which gives notes that are nearly pure tones. The music sounds woolly and indefinite, not so much from the special quality of the notes, as from a feeling of want of definiteness in the intervals.

In the Mixture Stops of an organ each note has one or more additional pipes, which are tuned to the octave, twelfth, etc. of the note. Such a stop cannot well be used alone, but is most useful with the Full Organ, and especially in accompanying congregational singing, for by strengthening some of the harmonics it accentuates the relationships of the notes to each other, and makes it easier for the singers to keep in tune.

271. Advantages of the diatonic scale. In Chapter I we described the musical scale known as the True or Diatonic Scale, and gave the vibration ratios for the intervals included in the scale. This scale, or a scale approximating to it, has been in use amongst European nations for many centuries, and its origin is unknown. We have seen that the intervals which are found in the scale can be obtained by the subdivision of a string, and it is possible that the scale took its rise from this fact, for the ancients were acquainted with the natural tones of vibrating strings. Whatever may have been its origin, the scale has survived to the present day, and there can be no doubt that the reason for its survival is that no other scale is so well fitted to provide harmonious combinations.

The following method of deriving the scale by combining the most consonant intervals shews its advantages from the point of view of harmony, but it is not suggested that it took its rise from any such considerations.

Suppose we wish to build up the scale on the note C.

Add to C the notes which make the best consonances with it, namely, the Fourth, the Fifth, and the Octave, and call these F, G and *c* respectively. We have now the following notes in ascending order of pitch:

$$\begin{array}{cccc} C & F & G & c \\ 1 & \dfrac{4}{3} & \dfrac{3}{2} & 2. \end{array}$$

If the frequency of the lowest note is taken as unity, the frequencies of the other notes will be given by the numbers placed below them.

We saw that the best combination of three notes is the Major Triad. Let us then place a Major Triad on each of the notes C, F and G, and see what new notes are introduced. It does not matter which form of the Major Triad we use. The result will be the same whether we use the fundamental form or either of its inversions. Taking the fundamental form, and remembering that the frequencies of the three notes are in the ratio of 4:5:6, we get for the triad on C the frequencies 1, 5/4 and 3/2. For the triad on F we get 4/3, 5/3 and 2, and for the triad on G we get 3/2, 15/8 and 9/4.

Now arrange all these notes in order of frequency, bringing the note 9/4 down an octave, so as to bring it between C and *c*. We have then the following scale:

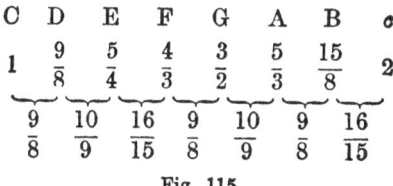

Fig. 115

and this is the ordinary diatonic scale. The note C is called the tonic of the scale, and the fraction immediately below any note is the vibration ratio of the interval formed by that note and the tonic. The lower row of fractions is got by dividing each fraction in the upper row by that on its left, and gives the vibration ratios of the intervals between the pairs of consecutive notes of the scale.

This scale, modified somewhat to meet the exigencies of modern music and instruments, is in almost universal use by civilized nations. One or two other scales, such as that of the Scotch Bagpipes, have survived from former times, but they are quite exceptional, and perhaps owe such charms as they possess to the contrast they present to the superior sweetness of the more familiar scale.

So long as music is used only for melody, it is of no great consequence what scale is used. Some music-loving though half-civilized nations have in fact no fixed scale at all. Their vocal music is only an accentuated form of speaking, the voice rising and falling not by fixed steps, but by a continuous glide. When once harmony is introduced, a definite scale is required, and the diatonic scale was generally adopted long before Helmholtz shewed why it is so much better fitted than any other to provide consonant harmonies.

The intervals between consecutive pairs of notes are of three kinds; the Major Tone 9/8, the Minor Tone 10/9 and the Semitone 16/15, and the intervals must be arranged in the order shewn in Fig. 115 to give a true diatonic scale.

Each of the notes has received a name which indicates its relationship to the tonic. These names are as follows:

$$\begin{array}{ll} \text{C Tonic} & \text{G Dominant} \\ \text{D Supertonic} & \text{A Submediant} \\ \text{E Mediant} & \text{B Leading Note} \\ \text{F Subdominant} & \text{c Octave.} \end{array}$$

These names are given not to notes of specified absolute pitch, but to notes making specified intervals with the tonic. If a diatonic scale is built up on the note D as tonic, the subdominant, for instance, will now be G, a fourth above D.

272. Modulation. Defects of the Diatonic Scale. A characteristic of modern music is that it frequently modulates or changes its tonic. A composition may begin with C as tonic, and presently change into the key of G; that is to say it now requires a diatonic scale with G as the tonic. Let us find whether the diatonic scale of C, when extended in both directions by raising or lowering all its notes by one or more octaves, will provide the notes required for a diatonic scale with G as tonic. Keeping to the scale of frequencies in which the frequency of C is represented by unity, we have 3/2 for the frequency of G, and this must be multiplied by each of the fractions in the middle line of Fig. 115 in turn to give the frequencies of the notes required for the Diatonic scale of G. Without needing to perform the multiplications, we can see at once that some extra notes will be needed. The interval from tonic to supertonic is 9/8, whereas the interval from G to A is 10/9, which is a smaller interval than 9/8. Hence for the supertonic of G we need a new note a little sharper than A. The note B will serve as the mediant of G, since it is a true Major Third above G. Also c will serve as the subdominant, as it is a fourth above G. Similarly, if D and E are raised an octave, they will serve as the dominant and submediant of G; but F is much too flat to serve as the leading note. It should be 9/8 above the submediant, whereas it is only 16/15 above.

Thus the transposition to G as tonic requires the addition of two new notes to the scale. Similarly to give a true diatonic scale on F as tonic we shall require again two new notes, one of them a semitone above A, and the other a little flatter than D. If we proceed in the same way to take all the other notes in the scale of C as tonics, we shall introduce other new notes. It is found that eleven such notes have to be added to those belonging to the key of C in order to provide a diatonic scale with each note of the key of C as tonic. Nor is this the end of the matter. The Minor Keys have to be provided for, and this again requires additional notes, for the Minor Scale of C for instance requires a minor third, a minor sixth and a minor seventh. Further, modern music often requires "accidental" notes, a semitone above or below the notes of the scale, and the same note will not serve both for C♯ and D♭, for instance, for two semitones of the diatonic scale do not make a tone. Moreover it must be possible to take any one of these accidentals as the tonic of a diatonic scale.

Thus it will be seen that to provide for all these contingencies a very large number of notes will be required in each octave. This causes no difficulty with unaccompanied vocal music, for the voice can adjust its pitch so as to give

246   DEFINITION OF INTERVALS   [CH. XV

true intonation in any key, and the same is true of such instruments as violins and trombones, which have no fixed keys, the pitch of each note being determined by the performer. Instruments such as the pianoforte, the organ, and most of the orchestral wind instruments, which have fixed keys giving notes of definite pitches, are in a different position. The performers on such instruments cannot adjust the pitch of each note to true intonation. They must take the pitch provided for them by the instrument maker or tuner, and it is plainly impracticable to have a great number of keys in each octave.

273. Temperament. Consequently a compromise has to be effected. The notes in each octave are limited to such a number as is found practicable, and some of the intervals are altered a little from the true diatonic intervals, so as to make it possible to modulate without departing greatly from the diatonic scale.

The number of notes to the octave is twelve for all instruments in ordinary use at the present time. On the pianoforte the white keys give a scale not much different from the diatonic scale, and five black keys are added, which give notes dividing the intervals of a tone into two semitones, differing a little from the true diatonic semitones.

In discussing methods of tuning it is convenient to make use of a smaller interval than any we have employed hitherto. The interval generally used is called the *Comma*, and is defined as the difference between the Major Tone 9/8 and the Minor Tone 10/9. Its vibration ratio is therefore $9/8 \div 10/9$ or $81/80$. This is about a fifth part of a semitone. A still smaller interval named the *Cent* is also often used. The cent is the 1200th part of an octave, or, since there are on the usual system of tuning 12 semitones in an octave, it is the 100th part of a semitone.

Only two methods of tuning, or *Temperaments*, as they are called, have been used at all extensively, and one of these is now practically extinct.

274. Mean-tone Temperament. The Mean-Tone Temperament was in common use in organs until 50 years ago. As it has now been displaced by Equal Temperament, it is not necessary to give any lengthy account of it.

If we tune upwards four true fifths from C, we reach a note which is a comma above the true E. Each rise of a fifth increases the vibration ratio by the factor 3/2. Hence if the frequency of C is called unity, that of a note four Fifths above C is $1 \times 3/2 \times 3/2 \times 3/2 \times 3/2$ or 81/16. To reach the note E which is nearest to the note four Fifths above C we must rise two octaves and a major third. Hence the frequency of this note E is $1 \times 2/1 \times 2/1 \times 5/4$ or 20/4. We find then that the interval between the untrue E and the true E is $81/16 \div 20/4$ or 81/80, which is a Comma. The untrue E can be made to coincide with the true E by reducing each of the fifths by a quarter of a comma, and this flattened fifth is the basis of the Mean Tone system. Briefly stated, the scale is obtained by rising or falling repeatedly by two of the flattened fifths, and returning by a true octave, until a sufficient number of notes have been obtained within the compass of an octave. The main feature of the scale is that, if only keys not far removed from C are used, such as G, F, B♭, D, the fifths are all a quarter of a comma flat, the major thirds are true and the minor thirds are a quarter of a comma flat.

These divergences from true intonation are not very noticeable, and therefore the scale has a good effect in these keys. When however we modulate into keys such as F♯ or B, which are remote from C, in that they require the use of many of the black keys of the pianoforte, the intonation is so untrue, that such keys are called "wolves," and cannot be used. The Temperament receives its name from the fact that each tone in the scale is the same, and is the mean of the major and minor tones.

275. Equal Temperament. Modern music demands free access to all keys, and therefore the Mean-Tone Temperament, which permits of modulation into only a few keys, has given place to Equal Temperament. In this system the octave is divided into twelve equal intervals called *Equal Temperament Semitones*. It is clear, then, that whatever may be the defects of the scale of C, they will be exactly the same in the key of C♯ or any other key, for to reach any key we raise each note of the key of C by the same number of equal semitones. Thus all the keys are equally good or equally bad, and modulation makes no change in the consonance.

The Equal Temperament Semitone is an interval which gives an octave when added to itself twelve times. If therefore its vibration ratio is p/q, the fraction p/q multiplied by itself 12 times must make 2, or $(p/q)^{12} = 2$. This is easily solved by the use of logarithms, and it is found that p/q or $\sqrt[12]{2}$ is a little less than 1·06, or expressed as ratio it is nearly 106:100.

On the piano or any instrument tuned to Equal Temperament, twelve fifths make seven octaves. In true intonation the vibration ratio for an interval made up of twelve fifths is $(3/2)^{12}$, and that of an interval of seven octaves is $(2/1)^7$. The former of these is greater than the latter in the ratio of 531441 to 524288, and therefore the Equal Temperament fifth is flatter than the true fifth by one twelfth of the interval defined by this ratio. Expressed in terms of the comma the Equal Temperament fifth is one eleventh of a comma flat.

Similarly three Equal Temperament major thirds make an octave, whereas three true major thirds make the interval $(5/4)^3$ or $125/64$, which is less than an octave. Hence the Equal Temperament major third is sharper than the true major third. The difference is 7/11 comma.

The divergences of the Consonant Intervals in Equal Temperament from those in true intonation are shewn below.

Octave	True		
Minor Third	$\frac{8}{11}$	Comma	flat
Major ,,	$\frac{7}{11}$	,,	sharp
Fourth	$\frac{1}{11}$	,,	,,
Fifth	$\frac{1}{11}$	,,	flat
Minor Sixth	$\frac{7}{11}$	,,	,,
Major ,,	$\frac{8}{11}$	,,	sharp

It is seen then that in present-day music no intervals are true except the octaves; and the thirds and sixths differ quite conspicuously from the true intervals. Hence according to Helmholtz's theory the adoption of Equal Temperament must have introduced into music dissonance which would not be present if true intonation were used. This conclusion deserves a little consideration.

In the first place Equal Temperament tampers with just those intervals which can best stand it. The octave is very closely hedged in by harmonics, which beat strongly, if there is mistuning. Consequently, every Temperament is compelled to keep the octaves true. A mistuned octave would not be tolerated in music. The fifths are the next in order of closeness of definition, and in Equal Temperament they are only 1/11 comma flat. The beats due to this mistuning are not so rapid as to be seriously unpleasant. It is the thirds and sixths which suffer most in Equal Temperament, and we saw that these intervals are only feebly defined, so that we can tolerate more mistuning in their case than we could in the case of the fourths and fifths.

Helmholtz's Theory is a *physical* theory which aims at explaining the physical property of smoothness of concords, and does not necessarily arrange the concords in their order of desirability from an aesthetic point of view. The major third on his theory is much inferior to the fifth or octave; yet musically it is more agreeable. The absence of a third in a chord makes it thin and uninteresting, and it is a general though not invariable rule in Harmony to include a third in every chord. Moreover, actual dissonances such as the major and minor ninth are often introduced with excellent effect. Thus we may conclude that smoothness is not the only desirable feature of a chord in music.

Our appreciation of Tempered Intonation is no doubt largely a matter of education. As Donkin says, "the whole structure of modern music is founded on the possibility of educating the ear not merely to tolerate or ignore, but even in some degree to take pleasure in slight deviations from the perfection of the diatonic scale."

CHAPTER XVI

MUSICAL INSTRUMENTS

276. Classification of Orchestral Instruments.
We shall not attempt to give a complete account of the construction and use of musical instruments, but shall merely touch on such points as afford illustrations of the Acoustical Theories developed in the preceding chapters.

The instruments in common use in the Orchestra may be divided into four main classes, Strings, Wood-Wind, Brass, and Percussion. We shall treat them in this order.

The Strings may be again subdivided into (1) the Pianoforte and Harp, where the vibrations are produced by striking or plucking the strings and (2) the Violin class, where the strings are bowed.

277. The Pianoforte. In the pianoforte the strings are stretched on a wooden or metal frame, and a thin wooden sound-board is fixed to the frame. As has been said previously, if the strings were fastened to a rigid frame without a sound-board, very little sound would be given out. The sound-board must not be regarded as a resonator. Its vibrations are forced and it owes its efficiency merely to its large surface. It has of course natural tones of its own, but the damping is so great that after a very few vibrations its natural vibrations become inappreciable, and nothing remains but the forced vibrations.

Each note has from one to three strings. When there are more than one, the strings are tuned in unison with each other. We saw in Chapter III that the pitch of a string can be altered by changing its mass, length or tension and all three methods are employed in the pianoforte. The bass strings are several feet long, whilst those at the treble end are only two or three inches long. Further, the bass strings are weighted by one or more layers of wire twisted round them. This is better than merely using a thicker string, as it does not interfere so much with the flexibility. The strings are made longer and heavier in the bass in order to equalize the tension. If the strings were of the same length and density throughout, the tension in the treble would have to be much greater than in the bass. A pianoforte has generally a range of about 7 octaves. With this range the highest note has a frequency 128 times as great as the lowest, and if this were to be secured merely by difference of tension the highest string would have 16,384 times the tension of the lowest. This is an impracticable range, for, even though the upper strings had such a tension that they were on the point of breaking, the lower strings would be so slack that they would give a very poor and weak note.

The strings are struck with hammers covered with felt. We saw that, when a string is struck, no harmonic is present that requires a node at the point struck, and also that a string gives a more metallic tone when struck near the end than when struck near the middle. It has often been stated that pianoforte strings are struck one seventh of their length from the end in order to get rid of the seventh harmonic, which is the lowest that falls out of the musical scale. It is in practice seldom that a string is struck so far from its end as one seventh; a more usual position is one eighth or one ninth from the end. The treble strings are struck still nearer the end, for their stiffness hinders the formation of the higher harmonics, and their quality would be different from that of the lower strings, if the production of the higher harmonics were not encouraged by the nearness of the point struck to the end.

The quality of the note is also influenced by the shape and hardness of the hammer. A hard hammer and a narrow striking surface both favour the production of high harmonics; consequently narrower hammers are used for the treble in order to equalize the quality.

Each maker adopts the point of striking and the kind of hammer that he has found by experience give the quality he desires. The differences between the instruments of different makers depend largely on the choice of hammers and striking points.

278. The Harp. Much of what has been said of the pianoforte applies also to the harp. The bass strings are weighted, so that the tensions may be equalized. The soundboard is much smaller than in the pianoforte and the sound is consequently weaker. The strings are usually plucked not far from their middle points and the note is therefore soft and somewhat dull from the weakness of the higher harmonics.

*279. Motion of a Violin string. The Vibration Microscope. The Violin, Viola, Violoncello and Double Bass are alike in principle, differing merely in size. In all these instruments the string is made to vibrate by being bowed. The motion of a violin string under the action of the bow has been investigated experimentally by Helmholtz by the use of the Vibration Microscope. Briefly stated, the principle of the vibration microscope is as follows. A tuning-fork, whose pitch is the same as that of the string, has a lens fixed to one of its prongs, and is placed so that the vibrations of the lens and those of the string are in directions at right angles to each other. A grain of starch is fastened to the string. If now the fork is made to vibrate, the image of the grain will vibrate, and a person looking at the grain through the lens will see it drawn out into a short line. If the string is vibrating and the fork at rest, the grain will be seen drawn out into a line at right angles to the former. If both fork and string vibrate the two separate motions of the image are compounded with each other, and a Lissajous' Figure is seen. The vibrations of the lens are simple harmonic, those of the string are periodic but not simple harmonic. Consequently the figure will not resemble any of those shewn in Chapter II, since those figures were drawn for the case in which both vibrations are simple harmonic.

The general form of the figures seen by Helmholtz was as shewn by the curve at the top of Fig. 116. Here the fork is describing simple harmonic vibrations in a horizontal direction, and the string is vibrating in a vertical direction, the fork and string being in unison. It remains to deduce the relation between the displacement of the string and the time. This is readily done, as the known mode of vibration of the fork provides us with a time scale in the horizontal direction. At a certain moment, for instance, the displacement of the string at the point we are observing is ab upwards, and the moment at which the displacement has that value can be determined from the displacement oa of the fork at the same moment.

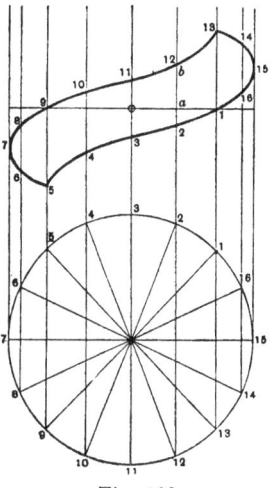

Fig. 116

Draw a circle with its centre anywhere in the vertical line through O, the centre of the Lissajous' Figure, and with a radius equal to the amplitude of vibration of the fork. Divide the circumference of the circle into any number of equal parts, and through each of the points of division draw a vertical line. In the figure there are 16 sections, the points of division being numbered from 1 to 16. The point numbered 1 is for convenience chosen so as to lie vertically under one of the points where the Lissajous' Figure cuts the horizontal line through O. The vertical lines will mark a scale of equal time intervals on the horizontal line through O. The numbers on the curve shew the positions of the tracing point of light at the ends of successive equal intervals of time.

Now take a straight line divided into 16 equal parts as in Fig. 117, and at each dividing point draw an ordinate equal to the ordinate of the Lissajous' Figure at the point with a corresponding number, and draw a smooth curve through the ends of the ordinates. The curve is found to consist of straight lines meeting at angles with each other.

This curve does not shew the shape of the string, but the displacement of one particular point of the string at different times. The tangent of the angle between the curve and the axis at any point is the ratio of a change of displacement to the time interval in which that change takes place, or is the velocity of the point of the string. Consequently a straight line represents uniform velocity, and we see that any point of the string moves with uniform velocity downwards, then changes its direction of motion suddenly and moves with uniform velocity upwards, and so on. The to and fro velocities are the same only at the centre of the string; they

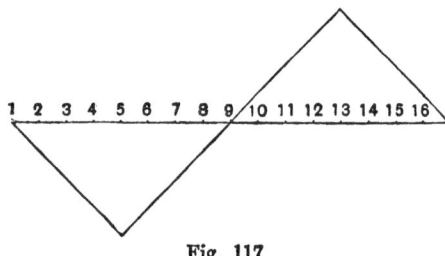

Fig. 117

differ from each other the more, the nearer the point observed is to the end of the string. Helmholtz found that at any moment the string takes the form of two straight lines meeting at an angle. The angle is not always at the same point of the string, but travels back and forwards along a flat curve, which passes through the ends of the string. When the angle is travelling in one direction it is above the equilibrium position of the string, and when travelling in the other direction it is below it.

280. Action of the Violin Bow. The action of the bow in exciting vibrations in the string depends on the difference between static and kinetic friction. When the string and bow are at rest relatively to each other, the friction is greater than when there is relative motion. The bow moves with uniform velocity, and carries the string forward with the same velocity. Presently the force of restitution becomes so great that the string breaks away from the bow, and Helmholtz's experiment shews that it returns also with uniform velocity. When it reaches the end of its swing and stops, another part of the bow grips it, and carries it forward again, and so on.

281. Quality of the note of the Violin. It will be seen that the vibrations are not at all like simple harmonic vibrations. The string does not slow down gradually as it reaches the end of its swing, but changes suddenly from an outward uniform velocity to an inward uniform velocity. A vibration of this kind requires a large number of terms of the Fourier series to express it, and the note of a violin has therefore a large retinue of harmonics. Helmholtz considers that the cutting character of the note is due to the strength of the sixth to the tenth harmonics as compared with those of other instruments. The nature of the vibrations is not much affected by the position of the point that is bowed.

282. Production of the Scale on a Violin. The lower strings of a violin are heavier than the higher strings, for the reasons given when we spoke of the pianoforte. The strings of the violin like those of all other stringed instruments are tuned by alteration of their tension. They are tuned to the notes g, d^1, a^1, e^2, making fifths with each other, and it is to be noticed that if the fifths are true, it is not possible to play an equally tempered scale on the instrument making use of the open strings, for all the fifths on the tempered scale are $1/11$ comma flat. The point has merely a theoretical interest, for $1/11$ comma is too small an interval to be of practical consequence.

The notes intermediate between those to which the strings are tuned are obtained by pressing the strings against the finger-board with the fingers, and so shortening the strings by the required amount. On some stringed instruments, such as the banjo, small raised strips of metal or ivory called Frets are fixed across the finger-board at the points to which the strings are to be shortened for the various notes of the scale. These frets make the instrument easier to play, but they have the disadvantage that if one of the strings is out of tune, all the notes produced from that string must be out of tune, whereas if one string is out of tune on the violin, the player can adjust the point of stopping, so as to bring all except the

256   MUSICAL INSTRUMENTS   [CH. XVI

open note into tune, and the note that should be given by the open string can be obtained from the string below.

The peculiar shape of the body of the violin is beyond the reach of theory. It was arrived at by experience and has remained practically unchanged for 200 years.

283. The Wood-Wind Instruments. The Wood-Wind consists of two classes. The first class contains the Flutes and the second contains the Reed Instruments, such as the Hautboy, Clarinet and Bassoon.

284. The Flute. The Flute is made in various sizes, the best known varieties being the orchestral flute, the military flute and fife, and the piccolo. The piccolo is the instrument of the highest pitch used in music. Each of these instruments consists of a tube closed at one end. In the side of the tube near the closed end is a hole across which the player sends a sheet of air from his lips and so produces the sound, the vibrations being set up in the same way as in the flue pipes of an organ. The flute then is analogous to an open organ pipe, and gives the full series of harmonic overtones. Flutes were formerly made with the bore of the half of the tube nearest to the open end slightly conical, the narrowest part of the cone being at the open end. The conical bore is still used in military flutes, but orchestral flutes are now always made with a cylindrical bore. The bore near the mouth hole is commonly slightly contracted in the form of a paraboloid.

285. The Finger Holes. In the half of the tube farthest from the mouth are six holes which are used for forming the scale. In the so-called eight-keyed concert flute these holes were covered with the fingers. In the Boehm flute and others of modern make they are covered by padded keys, which are pressed down with the fingers. When all the holes are closed, the flute gives the note of an open pipe whose length is the distance between the mouth-hole and the open end. If the holes were as large as the bore of the tube, they would reduce the effective length of the pipe to the distance between the mouth-hole and the highest hole left open, and the distances of the holes from the mouth would have to be inversely proportional to the frequencies of the notes of the scale. It is not practicable to make the holes so large as this. If they are above a certain size they are not easily covered, and the notes produced are unmanageable. In the case of the eight-keyed concert flute the holes were much smaller than the bore. When the holes are small the notes of the instrument are weak, and it is largely for this reason that in modern flutes the holes are covered by keys, and can therefore be made larger than is possible when they are covered by the fingers. It is claimed also for the Boehm and other similar flutes that the holes can be placed more nearly in their theoretically correct positions, as the fingers need not be directly over the holes, but can open and close them by means of levers.

A simple experiment will shew that the part of the tube below the highest hole open is not without effect on the pitch of the note. The experiment can be made with a tin whistle. Cover the two highest holes and blow the whistle. Now cover also the three lowest. The highest hole open is the same as before, yet the pitch is lowered a little. Thus it appears that the tube is not completely cut off at the highest open hole. A warning should be given that the experiment will not generally succeed if the highest hole is left open and the remaining five alternately opened and closed. In this case closing the lower holes can with great care in blowing be made to lower the pitch, but it will generally raise it a semitone. When all the holes are open, the pipe between the mouth and the highest hole gives its fundamental. When all the holes except the highest are closed, we shall almost certainly get the first overtone of the full length of the pipe, which is a semitone above the note given out when all the holes are open. The first overtone has an antinode in the middle of the pipe, and a node a quarter of the length from each end. If all the holes are open, we cannot have a node in the lower half of the pipe. The only possible position for a node is in the closed upper half. When all the holes but the highest are closed, it is almost impossible to prevent the formation of an antinode at the middle of the pipe, since that point is connected with the open air through the highest hole. As it is now possible for a node to form in each half of the pipe, the first overtone of the whole pipe is produced.

286. The Flute regarded as a Resonator. As a first approximation we can regard the flute as a pipe whose length can be varied, but this will not explain the whole of the details of the construction. The holes are not arranged so that their distances from the mouthpiece are inversely proportional to the frequencies of the notes produced, and they are not always all of the same size.

We can get a step farther in the explanation by looking on the flute as to some extent analogous to a Helmholtz Resonator, whose pitch can be raised by enlarging the opening. Uncovering the holes of a flute is equivalent to enlarging the opening of a resonator, and the larger a hole is, the greater is its effect in raising the pitch. The part of the tube in which the holes are open is not quite freely open to the air, and is therefore not quite without effect on the pitch of the note given out. This explains why it is not a matter of indifference whether the holes below the highest open hole are open or closed. Closing them reduces the connexion of the whole interior with the open air, and so lowers the note. It is similar in effect to shading the open end of an open organ pipe.

We can now understand why the holes may, within limits, be made in any positions convenient for the fingering. If when so arranged they do not give a true scale, they can be altered in size so as to correct the errors.

With merely six holes as described, the flute will give a diatonic scale extending over about three octaves. For the lowest octave the fundamental of the pipe is used, for the second octave the first overtone an octave higher is used, and for the third octave the overtone two octaves above the fundamental is used. The semitones intermediate between the notes of the scale are produced by means of holes covered by keys, which can be opened when required.

An open hole prevents the formation of a node in its neighbourhood, but favours the formation of an antinode. This principle is made use of in the production of certain high overtones. The holes near the points where nodes are situated in the particular form of vibration required are closed, whilst the holes near the antinodes are left open.

The note of the flute is almost free from the higher harmonics. The octave is faintly audible, but no others. Consequently the note has a smooth quality which contrasts well with that of the violins and reed instruments.

287. The Ocarina. In an earlier chapter we mentioned an instrument called the Ocarina. It is not used in the orchestra, but is worth a short description, as it is a simple resonator in principle, with none of the characteristics of a pipe.

The Ocarina is a hollow pear-shaped instrument generally made of metal. *A* is a flat tube by which a sheet of air is

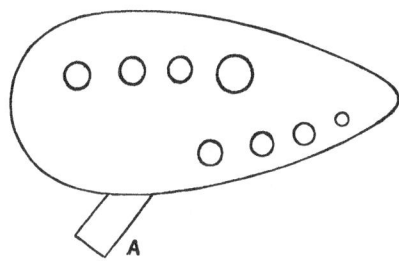

Fig. 118

blown across a hole not seen in the figure, and the instrument made to give out a note. On the front are eight holes which can be covered by the fingers, and on the back are two holes for the thumbs. The scale is produced by uncovering the holes one after the other. The interesting point about the instrument is that the positions of the holes are of no consequence. All that matters is their size or more strictly their conductivity. Choose two holes of the same size, and open first one and then the other. They will be found to have the same effect in raising the pitch wherever they are situated. The instrument has not a vibrating *column* of air with nodes and antinodes, but merely a *mass* of air which is alternately compressed and rarefied, and the pitch depends only on the volume of the air and the total conductivity of the openings.

288. The Clarinet. We come next to the reed instruments, comprising the Clarinet, Hautboy and Bassoon, with some others in less general use, such as the Basset Horn, Cor Anglais, Double Bassoon, Saxophone, etc.

The Clarinet has a cylindrical tube spreading out into a small bell at one end, and closed at the other end by a single reed of cane, beating on a rectangular opening in the side of the mouthpiece. We have already explained the action of a reed in Chapter X, and have shewn that the reed end of a pipe is to be treated as a closed end. The clarinet therefore, being a closed cylindrical pipe, gives only the odd harmonic overtones 1, 3, 5, etc. When overblown its note rises a twelfth, unlike that of the flute which rises an octave.

The notes of the scale are produced by opening in turn a series of holes in the side of the tube, but as it is necessary to cover a range of a twelfth before beginning again with the first overtone, keys are provided for giving a few notes below and above those obtained by the use of the holes. A clarinet in B♭ gives the note F when all the finger holes are closed, but E, E♭, and D can be got by the use of keys, which cover other holes below the lowest of the finger holes. Similarly when all the finger holes are open the note given out is F an octave above the former, and keys above the highest hole enable the player to produce F♯, G and G♯. Thus by the use of the keys and holes the tube, whilst always sounding its fundamental, can be altered in length so as to give a scale extending over a semitone less than a twelfth. Now return to the fingering used for the lowest D, but make the tube give its first overtone, and we get the note A a twelfth higher. We then get by repeating the former fingering a scale in which all the notes are a twelfth higher than in the lower register.

Throughout the upper register a key called the Speaker Key is used to facilitate the formation of the first overtone. In the lower register there is a node at the reed and an antinode somewhere near the highest hole that is open. In the higher register there is a second node between the antinode at the highest open hole and the reed, and a second antinode between this second node and the reed. The speaker key opens a small hole near this second antinode and encourages its formation. As a single speaker key has to serve throughout the register, it cannot be exactly at the antinode for every note of the register, but it is sufficiently near to ensure the production of the first overtone of the tube.

Most of what was said of the production of the scale on the flute applies also to the clarinet. The intermediate semitones are provided by means of additional keys, and the highest notes are obtained by a method similar to that employed on the Flute.

As the proper tones of the clarinet form the odd series of harmonics, none but the odd harmonics are conspicuous in its note, and this is the main cause of its characteristic quality. The even harmonics are not quite absent, for the periodic current of air from the reed contains the whole series, and the even members set up weak forced vibrations in the column of air.

The pitch of the clarinet depends almost entirely on the length of the tube. The reed has a natural frequency determined by its mass and elasticity, but its mass is so slight that it is constrained to vibrate with the frequency proper to the column of air.

Clarinets are made in a variety of pitches. Those in common use in the Orchestra are in A, B♭, C. The Basset Horn, a fourth below the B♭ Clarinet, and the Bass Clarinet, an octave below the B♭, are also sometimes used.

289. The Saxophone. The Saxophone is a brass instrument resembling the Clarinet in the form of its reed, but it has a conical tube, and therefore has the full series of harmonic overtones. Its fingering is similar to that of the Flute. It is seldom heard in this country, but is in common use in military bands on the Continent.

290. The Hautboy. The Hautboy has a conical bore, and the sound is produced by a double reed. Two thin pieces of cane slightly curved are bound together with their concave faces towards each other, so that there is a narrow lenticular opening between their edges, when they are at rest. When they are made to vibrate by the pressure of the air in the player's mouth, the lenticular space alternately opens and closes, thus admitting puffs of air into the tube. The reed is placed at the vertex of the cone formed by the bore of the tube, the base of the cone being at the open end of the instrument.

In consequence of its conical bore the hautboy gives the full harmonic series of overtones. Hence the arrangement of the holes and keys and the fingering are the same as for the flute. The notes of the hautboy are very penetrating in quality on account of the strength of their higher harmonics.

291. The Bassoon. The Bassoon is merely a Bass Hautboy. Its length is so great that for the convenience of the performer it is doubled on itself, and the reed is placed at the end of a short side tube. In consequence of the great length of the tube the holes are bored obliquely through the wood, so that whilst their outer ends are close enough together to be easily reached with the fingers, their inner ends are widely enough separated to give the notes required. In other respects the bassoon resembles the hautboy. The reed is double and the bore of the tube is conical. The overtones form the full harmonic series.

292. Tuning the Wood-Wind. The range within which it is possible to alter the pitch of an instrument of the wood-wind class to bring it to the pitch of a pianoforte or other instrument is very limited. The pitch of a flute, for instance, can be lowered by drawing out the joints a little, but this puts the notes of the scale out of tune with each other. The spacing of the holes is arranged to give the proper intervals when the tube has its normal length with all the joints pushed close. When the pitch is lowered, the section between every two holes should be lengthened in the same proportion as the whole tube is lengthened, if the relative dimensions of the instrument and the relative pitches of the notes are to remain unchanged. When the flute is flattened by drawing out the head joint, it is plain that the holes will be a little too close together to give the correct intervals with the increased length.

There is no difficulty in tuning the strings and brass by any amount that is likely to be needed and therefore it is usual to tune the orchestra to the wood-wind instruments, when the pitch is not fixed by the inclusion of an organ or pianoforte. The hautboy is generally chosen for this purpose, but probably the clarinet would be better. The flute and bassoon are what may be termed flexible instruments, that is to say, their notes can be modified a little in pitch by the player, and so brought into tune. The flute player adjusts the pitch by covering the mouth hole to a greater or less extent with his lips, and the bassoon player by varying the pressure on the reed. The hautboy and clarinet are less flexible, and perhaps the clarinet is the less flexible of the two. Consequently these instruments suffer most in intonation when their joints are drawn out, and it is for this reason that one of them is chosen to give the pitch to the rest of the orchestra.

293. Standards of Pitch. There are two Standard Pitches in use at the present time. Up to the year 1896 the High or Philharmonic Pitch had been in use for many years. This had a vibration number 452·4 for the note A at a temperature of 60° F. In 1896 the Philharmonic Society changed their standard to what is known as the Low Pitch, and has a vibration number 439 at 68° F. In specifying a standard of pitch it is necessary to specify the temperature, for the standard is fixed mainly for the guidance of the instrument makers, and instruments vary in pitch with variation of temperature. In fixing the Low Pitch as stated above, the intention is that the makers should construct the instruments in such a way that at a temperature 68° F. the note A shall have a vibration number 439. At any other temperature the instruments will not only have a different pitch but will also differ from each other, as different instruments vary in different ways with change of temperature, and they will therefore have to be brought into tune with each other by the use of the tuning appliances appropriate to the various instruments. The temperature was fixed at 68° F. as this is an average temperature for a concert room, and so entails the smallest amount of adjustment of the instruments. The Low Pitch has been adopted by all the great London Orchestras, but the High Pitch is still used in military bands and in most provincial orchestras. The difference between the two pitches is not great, yet it is enough to make it impossible to use the same clarinet or hautboy for both. Consequently, players who are in the habit of using both pitches are compelled to have two instruments, one for each pitch. If it were not for this difficulty, it is probable that the Low Pitch would now be universal in this country. Unfortunately the military bands have retained the High Pitch, as the expense of providing new instruments of Low Pitch is too great. As provincial orchestras are often dependent on the local military band for filling the gaps in their ranks, it would appear that we shall not have one standard of pitch in orchestras, until the military authorities can face the expense of providing new instruments.

294. The Brass Instruments. The Brass Instruments all consist of metal tubes of a more or less conical bore provided at the narrow end with a cup-shaped mouthpiece. The lips of the player are placed against the mouthpiece and by their vibrations produce the sound in the same way as the voice is produced by the vocal chords.

In all these instruments the aim of the maker is to provide the full series of harmonic overtones. The overtones form the basis of the scale as in the case of the wood-wind, but much higher members of the series are used. The horns and trumpets, for instance, go as high as the sixteenth harmonic.

295. Shape of the Brass Instruments. Though the bore is in general conical, it is by no means so regular a cone as the bore of a hautboy or a conical organ pipe. The bugle is a fairly regular cone, but the trombone is cylindrical for two-thirds of its length, and spreads only in the lowest third. All the instruments widen rapidly at the open end to form a bell. The positions of the nodes and the pitches of the overtones of such tubes cannot be calculated theoretically, and the makers have evolved by experiment the shapes that are found to give overtones in accordance with the harmonic series. Their efforts are not always successful, and the difference between a good instrument and a bad one is largely a difference in the accuracy of the pitch of the overtones. The nodes vary in position according to the overtone that is sounded. If there is a constriction in the tube at some point, any overtone which requires a node at that point is a little sharp, and any overtone which requires an antinode there is flat. A bad bruise in the tube may therefore have the effect of putting some of the notes out of tune.

296. Effect of shape on Pitch and Quality. As the shape of the bore of a brass instrument is not amenable to theoretical treatment, we can only mention a few experimental conclusions bearing on the relation of the shape to the quality and pitch of the notes that can be produced.

When the tube is wide relatively to its length, as in the euphonium, tuba, and other instruments of the Saxhorn group, the lower members of the harmonic series including the fundamental are easily produced, and are of good quality. When the bore is narrow, as in the horn and trumpet, the fundamental is difficult or impossible to blow, but the higher members are easy. The best range of the horn is from about the fourth to the twelfth harmonic.

A wide spreading bell, as in the horn, makes the tone smooth. A small bell, as in the trombone, conduces to a brighter quality. Widening the bell beyond the size for which the tangent at the edge makes an angle of about 45° with the axis has no effect on the pitch, though it continues to make the notes smoother in quality.

The shape of the mouthpiece has a great effect on the quality of the notes. A shallow cup-shaped mouthpiece like that of the trombone gives a bright tone. A deep conical mouthpiece narrowing gradually from the rim, like that used with the horn, gives a smoother tone.

297. The Bugle. Some instruments, such as the Bugle, Post Horn, and the French Cor de Chasse, have no other notes than those of the harmonic series. All the military Bugle Calls are formed from the following notes;
. Some higher notes are possible but difficult. The Regulation Bugle is in B♭, and therefore the actual sounds are all a tone lower than those shewn.

298. The Cor de Chasse and Hand Horn. The Cor de Chasse is similarly limited, but, the tube being long and narrow, the higher tones are more easily produced, and, as the harmonics lie closer together the higher we ascend in the series, the melodic possibilities of the Cor de Chasse are greater than those of the bugle.

The horn used in orchestras or, as it is often called, the French Horn, had until recently no mechanism for filling the gaps in the harmonic series. Its scale of open notes was therefore limited in the same way as that of the Cor de Chasse. The performer, however, placed his hand inside the bell, and by closing the opening to a greater or less extent could flatten each of the notes by a tone or more, and so produce a complete chromatic scale over a range of about two octaves from the third harmonic upwards. The notes so flattened were called stopped notes, and were greatly inferior in quality to the open notes. The horn was provided with a set of lengthening pieces or "crooks" by which it could be put into the key of the music to be played, the number of stopped notes required being thereby reduced. Horns of this kind are called Hand Horns. Horns are now always provided with valves, by means of which a complete chromatic scale is obtained over the whole range of the instrument without the use of stopped notes, as will be explained in § 301.

299. The Ophicleide Class. The earliest method in general use for filling the gaps between the successive harmonics was similar to that used in the wood-wind instruments. Holes, sometimes covered by keys, were made in the side of the tube, and by opening these in turn a scale was produced in the same way as on the flute.

The once popular Key Bugle was an instrument of this class, as was also the Ophicleide. Their chief defect was the inequality in the notes. A note that was produced with all the holes closed had the advantage of the softening effect of the bell, whilst for the other notes the vibrations did not extend to the bell, and the quality was less satisfactory. The last survivor of the class was the Ophicleide, which has now been superseded by the Tuba.

300. The Trombone. The second method of forming the scale is by the use of a slide, as used on the Trombone. There are three Trombones in general use, the Alto Trombone in E♭, the Tenor in B♭, and the Bass in G. They differ only in size and a description of the Tenor will serve for the three.

About two-thirds of the tube nearest to the mouthpiece is cylindrical, and the remaining third is conical. The cylindrical part is bent in the middle so that its two halves are parallel to each other and is made double, the outer part sliding telescopically over the inner. By drawing out the outer part the tube is lengthened in the same manner as the right-hand branch of the interference tube shewn in Fig. 57.

When the slide is closed, the Tenor Trombone gives the note B♭ as its fundamental. If the slide is drawn out a little way, the fundamental is lowered to A, if a little farther still to A♭, and so on, until with the greatest extension possible E♮ is reached. Thus there are seven positions of the slide giving all the semitones from B♭ down to E♮. In each of these positions we can produce not only the fundamental but also its harmonic overtones, and so we have a harmonic series on each of seven consecutive notes at intervals of a semitone.

The range of the trombone in practice extends upwards as far as the eighth harmonic. An expert player can produce a few higher notes, but they are seldom demanded by composers. In the following table are shewn all the notes that can be produced in the various positions of the slide. The fundamental is difficult to blow in any but the first three positions, and is therefore omitted in the rest. The seventh harmonic is omitted throughout, as it does not coincide with any note of the scale with the fundamental as tonic.

Fig. 119

It will be seen that from E♮, the second harmonic in the seventh position to B♭, the eighth harmonic in the first position, we have a complete chromatic scale, and three lower notes in addition. The slide has enabled us to bridge completely all the gaps above the second harmonic in the first position, and to bridge the upper half of the gap between the first and second harmonics. Some of the higher notes can be obtained in more than one position of the slide, and this is an advantage to the player. He chooses such positions as require the least movement of the slide in passing from note to note.

The trombones, like the violins, can be played in true intonation, since the pitch of each note is under the control of the player.

301. Valved Instruments. The third method of producing the scale on brass instruments is by the use of valves. The valve is a piston which, on being pressed down by the player, lowers the pitch of the instrument by throwing in an extra length of tube. There are generally three valves, the first of which lowers the pitch of the instrument two semitones, the second lowers it one semitone, and the third lowers it three semitones. As the valves can be used separately or together, they enable the player to lower the pitch by any number of semitones up to six, and thus answer the same purpose as the slide of a trombone.

If we have an instrument in B♭ provided with three valves, Fig. 119 can be made to represent the notes obtainable with the various combinations of valves in the following way:

Position 1 corresponds to no valve
,, 2 ,, valve 2
,, 3 ,, ,, 1
,, 4 ,, ,, 3 or 1 + 2
,, 5 ,, valves 2 + 3
,, 6 ,, ,, 1 + 3
,, 7 ,, ,, 1 + 2 + 3.

Here again the scale is complete from a note six semitones below the second harmonic. This is generally all that is needed, for on most brass instruments the fundamental is not used. The tuba and some other bass instruments of the Saxhorn class form exceptions. Their fundamental is of good quality, and these instruments have therefore a fourth valve, which lowers the pitch by a fourth or five semitones. With the help of this valve the pitch can be lowered by any number of semitones up to eleven, and therefore the gap between the fundamental and its octave can be bridged.

The introduction of valves has greatly increased the facility with which brass instruments can be played. Rapid passages can be played on them which would be quite impossible on a slide. They suffer however from the defect that, whenever two or more valves are used together, the note produced is a little sharp. The first valve, for instance, is tuned to lower the pitch a tone when used alone. This requires the length of the tube to be increased in the ratio of 8 to 9, or the valve must add to the tube one eighth of its length. Similarly the second valve used alone adds about one fifteenth to the length. If now the two valves are used together, the first increases the length by one eighth, but the second increases it by only one fifteenth of its original length, which is less than one fifteenth of its length when already increased by the first valve. Thus, though the two valves used separately give a true tone and semitone respectively, the two together give a lowering of pitch less than three semitones. In most cases the player corrects the error as well as he can with his lips. The error is not very great with three valves, and it is possible to force the note down a little by relaxing the pressure of the lips. The horn player is in a better position, for he can flatten the note by closing the tube a little more with his hand, whenever he uses two valves at once. The error is more serious in instruments with four valves, and in their case compensating valves are often used to correct it.

Brass instruments can often be put into several keys by the use of lengthening pieces or crooks. The Cornet, for instance, is generally in B♭, that is to say, it gives the note B♭ and its harmonics when no valves are used, but it is sometimes put into the key of A, by the insertion of a piece of tube between the mouthpiece and the instrument. If the valve tubes are adjusted to add the right lengths to the B♭

cornet, they will be too short for the A cornet, and must be readjusted, when a change is made from B♭ to A. Each of the valve tubes has a slide like that of a trombone, and it is with these slides that the adjustment is made.

All brass instruments used at the present day have valves with the exception of the trombones. Attempts have been made to introduce valves on these also, but without success— at least in this country. The solemn and dignified quality of tone of the trombone is not suited to rapid passages, and the gain in facility by the addition of valves does not make up for the loss in purity of intonation. Valve trombones are however common in Continental military bands.

302. The Trumpet. The Trumpet is the Treble representative of the trombone family. It had formerly a short slide by which the pitch could be lowered one or two semitones. The slide has now gone out of use, and its place has been taken by valves.

The Trumpet is normally in F and has crooks with which it can be put into any key down to B♭. Its tube is narrow in proportion to its length, and consequently it can be made to give the higher members of the harmonic series. The range within which it is most commonly used in modern music is from the second to the twelfth members of the harmonic series. The cornet is often admitted into the orchestra as a substitute for the trumpet, and it will be of interest to compare the two instruments.

303. Comparison of the Trumpet and Cornet. The Trumpet when lowered by a crook to B♭ has a tube about 9 ft. long, and its harmonics can be used as high as the twelfth. The Cornet in B♭ has a wider tube than the Trumpet, and is half the length. In consequence of its wider bore the harmonics above the sixth are difficult to produce. It follows that the pitch of the highest practicable note is about the same on the two instruments. The actual sounds of the open notes which can be easily produced are shewn in Fig. 120, the

Fig. 120

first series being the notes of the trumpet and the second those of the cornet. The fundamental is omitted in each case, as it is not used. The seventh and eleventh are enclosed in brackets, as they are out of tune.

Both instruments give with the help of their valves a complete chromatic scale over a range of two octaves or more, but there is a great difference in the certainty with which the notes can be produced. The player presses his lips more tightly together the higher he wishes to rise in the scale, and in order to produce a given note the pressure has to be adjusted to suit that note. Suppose the note B♭ in the middle of the treble clef is to be sounded. The trumpet player has an open note a tone above the B♭, and another rather more than a tone below, and if he adjusts the pressure a little wrong, he gets C or A♭, when B♭ is wanted. The B♭ of the Cornet is not so closely hedged in by other open notes, the nearest being a major third above and a fourth below. Consequently, the cornet player does not need the same accuracy of pressure as the trumpet player, and is not so likely to blow a wrong note. This is the main reason why an inferior cornet player is less likely to cause a catastrophe than an inferior trumpet player. The note of the trumpet is so brilliant and piercing, that, even when played softly, it is easily heard through the rest of the Orchestra, and, if the player makes a mistake, everyone hears it.

In the past few years instruments with the same length of tube as the cornet but the shape of the trumpet have been gradually making their way into the orchestra, and they have almost displaced the orchestral trumpet in F. These new trumpets in B♭ have much of the certainty of the cornet and the brilliancy of the trumpet.

The Trumpet parts written by Bach and Handel are so high that they are almost impossible on modern trumpets, and a form of instrument called the Bach Trumpet has been devised specially for these parts. The Bach Trumpet is merely a straight Coach Horn with two or three valves added.

The French Horn presents the same difficulties as the trumpet and for the same reason. Its best range is from the fourth to the twelfth harmonic, where the open notes lie close to each other. The bubbling sound made by the horn players when feeling about for their note at the beginning of a phrase is a not unfamiliar sound in the Orchestra.

304. The Saxhorn Class. The Cornet is the smallest member of a large family of instruments named Saxhorns from their inventor Saxe. The other members are the Tenor Saxhorn, the Baritone, the Euphonium, and the Tuba or E♭ Bombardon. The Tuba is the only member regularly used in the Orchestra. The Cornet and Euphonium are sometimes used, the former as a rule only when a Trumpet cannot be obtained.

The Saxhorns are essentially Military Band instruments. They are wide in bore, and therefore in the larger forms the fundamentals are easily blown, and are of good quality. From the facility with which the notes are produced they are capable of greater execution than any other brass instruments.

305. Tuning the Brass Instruments. The brass instruments are affected by temperature in the same way as are the flue pipes of an organ, but an additional complication is introduced by the breath and hands of the player. A small instrument such as the cornet is soon warmed throughout its length to nearly the temperature of the breath and is not greatly affected by changes in the temperature of the surrounding air. A large instrument such as the tuba contains such a great volume of air that it is not much affected by the breath of the player, and so is free to respond to changes in the temperature of the room. All the brass instruments have a short telescopic tuning slide, like the slide of a trombone, by which their pitch can be varied to bring them into tune with the rest of the orchestra. If the pitch were lowered considerably by the use of the slide, it would be necessary to adjust also the tuning slide of the valves, in order that the additional length of tube thrown in by any one valve might continue to bear its right proportion to the total length of the instrument. In practice this is not necessary, for any small defect in intonation caused by the use of the tuning slide is easily corrected by a slight alteration in the pressure of the lips on the mouthpiece.

306. The Drums. The remaining class of instruments, the percussion instruments, present few features of interest that have not been already referred to in the preceding chapters. With the exception of the Kettle Drums they are not tuned to any particular notes, and are merely used to mark the rhythm. The Kettle Drum consists of a hemispherical shell of metal with a skin of parchment stretched over its open end. The tension of the parchment can be varied by a number of screws distributed round the rim, and thus the pitch of the note can be altered. The practicable range of pitch for any one drum is a fifth. If the parchment is too slack the tone is bad, and if it is too tight there is a risk of its being torn. It is most usual to have two drums tuned to the tonic and dominant of the music, but other numbers of drums and other notes are often used.

CHAPTER XVII

APPLICATION OF ACOUSTICAL PRINCIPLES TO MILITARY PURPOSES

307. Detection of Aircraft by Sound. During the war of 1914–18 it became a matter of importance to find a method of locating Aircraft during the night or in foggy weather. For this purpose use was made of the sound of the engine, propellers, etc.

The unaided ears are able to judge approximately the direction from which a sound comes, but not accurately enough for military purposes.

Our power of locating a sound arises from our having two ears. It cannot depend to any great extent on the difference of intensity at the two ears, for ordinary sounds have too great a wave length to permit so small an obstacle as the head to cast a sound shadow (see § 127); and the difference of distance of the source from the two ears is too small to have much effect on the intensity, unless the source is very close to the listener. The difference of phase at the two ears is however not small, and it is probable that the ears have the power of detecting differences of phase, and of using these differences to judge the direction from which the sound comes. The difference of phase is greatest when the sound comes from one side of the listener, and it disappears when the sound comes from the front or back. In intermediate directions there is an intermediate difference of phase. If a source of sound is straight in front or straight behind, it is not easy for the listener to decide between the two positions, for in each case there is no difference of phase; but if the head is turned round a little the difficulty disappears. This is easily tested by closing the eyes and getting another person to make a sharp sound, such as that given out when two coins are snapped together.

307–308] APPLICATION OF ACOUSTICAL PRINCIPLES 275

Our having two eyes enables us to estimate the distance of objects by the stereoscopic effect, and this power is increased if the eyes are in effect placed wider apart, as in Prismatic Binoculars and Range Finders. Should we then in a similar way make our power of locating sounds more exact, if we could in effect increase the distance between our ears, and so increase the difference of phase?

Such a separation of the ears was used during the war for the location of Aircraft.

Two large conical trumpets are fixed with their axes parallel on a framework which can be rotated round a vertical axis. The trumpets are a few feet apart, and the small ends are connected by rubber tubes to the ears of the observer. If the trumpets with their axes horizontal are kept stationary and a source of sound, such as a man beating a drum, moves past them, say from the right to the left of the listener, the effect on the listener is found to be that the drummer appears to be straight out on the right until he is nearly in front of the trumpets, and then he seems to move rapidly across, until he is straight out on the left. The moment when he is exactly in front can be determined with considerable accuracy.

In locating an Aeroplane the trumpets are swept round on their vertical axis, until the listener judges that they are pointing to the source of sound. A very small movement of the trumpets from this position makes the sound seem to move far to the left or right, and it is found that the point of the compass from which the sound comes can be found very closely.

This method cannot give more than the direction in which the sound is travelling when it reaches the observer. The source of the sound is not necessarily in that direction, for there may have been refraction or reflexion of the sound waves.

308. Application to Mining. It sometimes happens that when a subterranean tunnel is being driven towards the enemy's lines, the enemy is at the same time heard to be driving a tunnel towards our lines, and it is important to know where he is working and whether there is risk of the two tunnels meeting. In this case also the origin of the sound can be located by bringing it simultaneously to the two ears by rubber tubes connected to two instruments named Geophones.

276 APPLICATION OF ACOUSTICAL PRINCIPLES [CH. XVII

The Geophone consists of a small circular wooden box containing a mass of mercury enclosed between two mica discs. Each of these discs has adjoining it on the side away from the mercury a small air space, which can be connected to the ear by a rubber tube.

If the Geophone is to be used merely for detecting the sound of mining anywhere near, without locating the source, the two air spaces of a single Geophone are connected separately to the two ears by the rubber tubes. The instrument is then laid on the ground. If sound vibrations are passing, the inertia of the mercury prevents any great movement of the mica discs, though the box itself is vibrating. Hence there will be periodic changes of volume of the two air spaces, and waves of sound will pass along the tubes to the ear.

If information is wanted as to the direction from which the sound is coming, two Geophones are used. One air space of each is closed with a cork or otherwise, and the two remaining air spaces are connected to the ears. The Geophones are then moved about on the ground until the sound seems to come from straight in front or straight behind. The source of sound must then be in a direction perpendicular to the line joining the two Geophones. There is in this case no difficulty in distinguishing between the two directions from which the sound might come, as the enemy is known to be in front.

309. Location of Sounds under Water. For some years before the war signalling under water had been used to a limited extent, the sound being received by a submerged microphone similar to a receiving telephone. A great impetus was given to the method by the prevalence of enemy submarines round our coasts during the war, and several appliances were constructed which not only detected the presence of a submarine, but also determined its direction from the observing ship.

Two methods were in regular use. One of the methods employed two submerged microphones connected to two telephones, one of which was held to each ear, and the direction from which the sound came was determined as in the case of the location of Aircraft as described in § 307. This method is not very satisfactory, as the microphones and telephones have definite periods of vibration of their own, and consequently give out most strongly the vibrations which happen to come near in period to their own proper tones, which alters the character of the sound so completely that location is difficult. A better result is obtained by using instead of microphones two simple small rubber chambers connected to the ears by tubes. Such chambers give fainter sounds but they have no proper periods and so transmit the sound more faithfully.

The second method uses a different principle. A thin sheet of metal framed by a heavy ring has a small microphone at its centre and is hung in the sea. If the metal disc is caused to vibrate, sounds will be heard in a telephone connected with the microphone. If the plane of the disc faces the direction from which the sound waves come, the sound is heard in the telephone, whilst if the disc is turned round so that the sound waves travel in a direction parallel to its plane, they will arrive at the two sides in the same phase, and no vibrations will be produced. If then the disc is suspended in the water with its plane vertical and is turned round to face various points of the compass, when a submarine is in the neighbourhood there will be a position in which the sound of the submarine is not heard and another position at right angles to the first where the sound is a maximum. In the former position the source of sound must be in one or other of the directions in which a horizontal line in the plane of the disc points, but the observations will not distinguish between the two directions. In order to get over this difficulty one side of the disc is screened by a baffle plate of some non-resonant material. The sound is then heard most distinctly when the unscreened side of the disc faces the source of sound.

310. Sound Ranging. A very successful application of Acoustical Principles to military needs is that of finding the exact position of an enemy gun from the sound of its discharge.

Imagine a gun at G, and observers at A, B and C who note the exact moment when the sound reaches them. Suppose A is the nearest of the three observers to the gun. Draw a circle with G as centre and GA as radius.

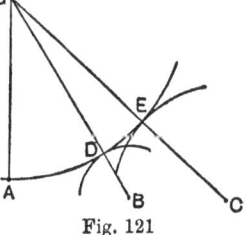

Fig. 121

278   APPLICATION OF ACOUSTICAL PRINCIPLES   [CH. XVII

ADE then will be the section of the wave front by a horizontal plane at the moment when the sound reaches *A*. The sound will reach *B* later than it reaches *A* by the time it takes to travel the distance *DB*, and it will reach *C* later than it reaches *A* by the time it takes to travel the distance *EC*. As the velocity of sound is known—correction being made for the temperature and humidity of the air—the observed delay of the sound in reaching *B* and *C* as compared with *A* can be converted into distances. Hence if we wish to work backwards to find the position of the gun from the observed retardation of the sound, we calculate how far the wave front was from *B* and *C* when the sound reached *A*, draw circles round *B* and *C* with these calculated distances as radii, and the gun will be at the centre of a circle which passes through *A* and touches the circles whose centres are *B* and *C*. The construction can be made on a map and thus the position of the gun will be found.

This method of finding the position of the gun is not simple geometrically, but the problem can be much simplified if it be assumed—as is the case in practice—that the distance of the gun from the stations *A*, *B* and *C* is much greater than the distances of *A*, *B* and *C* from each other.

Consider first only *A* and *B*. We know from the observations that *G* is at some point such that *GB* is greater than *GA* by a known amount. Hence *G* must lie somewhere on a hyperbola whose foci are *A* and *B*, and the direction of whose asymptotes is determined by the observed difference in the distances of the gun from *A* and *B*.

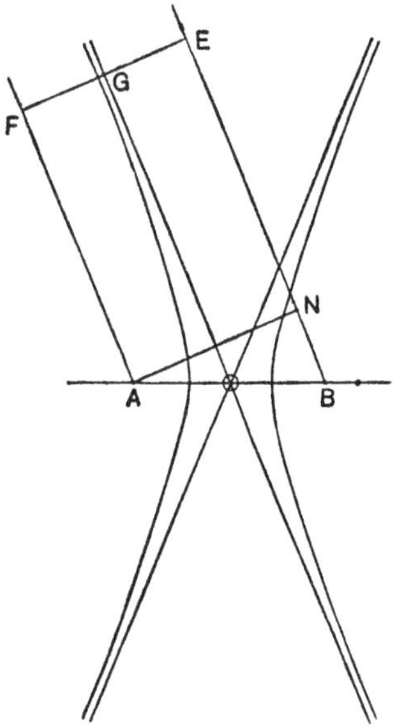

Fig. 122

Let Fig. 122 represent the complete hyperbola which is the

locus of points the difference of whose distances from A and B is the radius of the circle round B in Fig. 121. It is known on which side of the line AB the gun lies, and it is known that the gun is farther from B than from A, hence it must be on the upper half of the left hand branch of the hyperbola.

If the gun is very distant from O, the curve and its asymptote GO will be nearly coincident, and therefore GO may be taken as the locus of G. Draw AF and BE parallel to the asymptote, and draw lines through A and G perpendicular to the asymptote, then if OG is much greater than AB the difference between GB and GA will not differ appreciably from the difference between EB and FA. That is to say

$$GB - GA = NB \text{ nearly.}$$

But $\dfrac{NB}{AB} = \cos ABN = \cos GOA$, which gives the inclination of the asymptote, since AB and BN are known. Further, the asymptote passes through the point O midway between A and B. Its position is therefore completely determined and it can be plotted on the map. In exactly the same way a locus can be plotted from the pair of stations B and C, and the intersection of the two loci gives the position of the gun. Nothing would be gained by plotting a third locus from the stations A and C, for it would not be independent of the other two, and must of necessity pass through the same point if the figure is accurately drawn.

It is clear that at least three observation points are needed to locate the gun. In practice more than three are used for various reasons. The observing points are connected electrically with a central station, where the moment of the arrival of the sound at each is recorded photographically on a moving strip. If only three stations were used, one or two of them might be put out of action by shells cutting the wires or doing other damage. If half a dozen are used, the risk of fewer than three being in working order is lessened.

Further, an increase in the number of stations not only increases the accuracy of the location of the gun, but also gives an indication of the degree of accuracy attained. Suppose there are four stations A, B, C and D. A line on which the gun lies is obtained from each of the pairs AB, BC and CD. If these lines were strictly accurate they would meet at a single point. This extreme accuracy cannot be expected. In general the points of intersection of the three lines will not be coincident, but will form a small triangle. The three corners of this triangle give three separate determinations of the position of the gun, and its probable position is within the triangle. The larger the triangle, the more uncertain is the location.

Similarly if there are more than four stations the lines will all pass through a small area and the extent of this area is an indication of the accuracy of the result.

In the field the process thus described would take too much time, therefore a plotting board is prepared with the calculations made in advance. The board has on it a map of the district with the observing stations marked. Consider only stations A and B. A string is pivoted at a point half way between A and B. Along the edge of the board are marked time intervals between the arrival of the sound at A and B, and each of these intervals is marked on the board at such a point that the straight line from that point to the pivot of the string is the locus of the gun corresponding to that particular time interval. Thus as soon as the actual time interval between A and B has been found from the photographic record, all that needs to be done is to stretch the string over the point on the scale which gives that particular interval, and the string then gives a line on which the gun lies. Similarly a second string is pivoted at the point midway between B and C and this has its own separate scale on the edge of the board, so that it can be set to give a second locus. The point of intersection of the two strings is the position of the gun.

The moving strip on which the times of arrival of the sound are recorded sometimes contains also a record of the burst of the shell. This enables the observer to locate the point where the shell fell in exactly the same way as he locates the gun. He then knows the range and the approximate muzzle velocity, and as the muzzle velocity of many German guns was known to us during the war, it was often possible to form an opinion of the calibre of the gun. This estimate is not very trustworthy, especially in the case of guns which do not use fixed ammunition, but if the point where the shell fell is communicated to the artillery officer, he can search for fragments and so determine the calibre from the curvature of the shell.

QUESTIONS

Many of the following questions are taken from Examination Papers set in the Universities of Cambridge, London and Dublin and in the National University of Ireland.

Questions below the line in each section are somewhat more difficult than those above the line.

CHAPTER I

1. Can sound be propagated through a vacuum? Describe an experiment bearing on this question.

2. What characteristics distinguish a musical sound from a noise?

3. Describe an experiment from which you conclude that sound is due to the vibrations of the sounding body. Do all vibrations give rise to audible sounds? Illustrate your answer by examples.

4. On what characteristics of an air-wave do the intensity, pitch and quality respectively of the corresponding sound depend?

5. What experimental evidence leads to the belief that sound is propagated by wave motion?

6. How would you prove experimentally that the musical interval between two notes can be measured by the ratio of the vibration numbers of the two notes?

7. Explain how to use the Disc Siren with circles of 40, 50, 60 and 80 holes respectively to find the vibration ratio of a major sixth, assuming that the interval between the note G and the note e above it is a major sixth.

8. Shew that the measure of the difference between two intervals is obtained by dividing the vibration ratio of the larger interval by the vibration ratio of the smaller.

9. By what interval do two intervals whose vibration ratios are $\frac{3}{2}$ and $\frac{5}{4}$ differ from each other? What are these intervals and what is the interval got by adding them together?

---

10. Shew that if the interval between two notes is measured by the logarithm of the ratio of the frequencies of the two notes, the sum of the measures of two intervals will be the measure of the sum of the intervals.

CHAPTER II

1. What is meant by a perfectly elastic body and by limits of elasticity? Distinguish between a soft solid and a very viscous liquid.

2. State clearly the properties a medium must have in order that it may serve as a sound carrier.

3. State Hooke's Law and mention a few cases in which the law holds. Does it hold for gases?

4. What is meant by a Simple Harmonic Vibration? How is the displacement related to the acceleration in such a vibration?

5. What is meant by Isochronism? Give instances of isochronous vibrations.

6. Shew that in the case of the Simple Harmonic Vibration of a small mass the potential energy is proportional to the square of the displacement of the mass from its equilibrium position.

7. Shew that if a radius of a circle rotates with uniform angular velocity, the foot of the perpendicular from its end on any fixed diameter will perform simple harmonic vibrations.

8. Prove that a mass supported at the end of a spiral spring will execute Simple Harmonic Vibrations when slightly displaced. Calculate the periodic time on the assumption that the mass of the spring may be neglected.

9. Explain how the variations of the displacement and the velocity respectively with the time for a body executing simple harmonic vibrations may be represented by Sine Curves. What is the relation of the two curves to each other as regards their phases?

10. Define the terms period, amplitude and phase of a vibrating particle. Shew how the difference between the phases of two particles vibrating with the same period can be expressed by an angle.

11. Prove that in the case of simple harmonic motion the period of vibration is equal to 2π divided by the square root of the acceleration of the vibrating body when it has unit displacement.

12. Find the period of vibration of a mass of 1 kgr. attached to a spiral spring of such stiffness that an extra load of 15 gm. produces an extension of 1 cm.

13. Describe and explain a method of adjusting the pitch of a tuning-fork.

QUESTIONS 283

14. Describe the changes that take place in the curve traced out by a point which has simultaneously two simple harmonic vibrations in directions at right angles, the periods of the two vibrations being nearly, but not quite, equal.

15. A particle vibrates harmonically with a period of 2 sec. Find its amplitude if its maximum velocity is 10 cm. per sec.

16. Find the total energy of the particle in Example 15, assuming its mass to be 20 gm.

17. A string two feet long has its ends fixed at two points one foot apart and in the same horizontal line. Another string is attached to the middle point of the first string and has a bob at its lower end. Shew that when this second string has certain definite lengths the bob will describe Lissajous' Figures when set swinging.

Find the length of the second string when the bob is capable of describing the 2 : 1 figure.

CHAPTER III

1. Shew that when a train of transverse waves is passing along a series of particles the time taken by the waves to travel one wave-length is equal to the period of vibration of a particle.

2. From the relation in Question 1 prove the equation $v = n\lambda$.

3. Shew graphically that when two similar trains of waves travel in opposite directions along a string, stationary vibrations are produced.

4. Construct a diagram shewing the result of compounding two trains of waves one of which has double the wave-length and double the amplitude of the other.

5. Assuming that the velocity of waves on a stretched string is $\sqrt{T/\rho}$, find the period of vibration of a string when giving out its fundamental note.

6. How would you cause a stretched wire to emit its different harmonic overtones? What relations have these overtones to the fundamental of the wire?

7. How may the velocity of a wave in a string be deduced from a knowledge of the frequency of the vibrations and the positions of the nodes?

8. Explain how two bridges should be placed in order to divide a stretched string 100 cm. long into three segments whose fundamental frequencies are in the ratio 1 : 2 : 3.

9. Explain how you would use the monochord to prove that the frequency of vibration of a stretched string is proportional to $\sqrt{T}$. A certain string has a frequency 100. Find its frequency when both its tension and length are doubled.

10. Two strings of the same material and 12 and 15 in. in length respectively are stretched over a sounding board. If the tensions of the strings are produced by weights of 64 lbs. and 36 lbs. respectively, what will be the interval between the notes produced when the strings are plucked?

11. Two strings of the same length give the same note, but the tension of one is double that of the other. Compare the masses of the strings.

12. Two wires one of aluminium and the other of steel, identical in shape and volume, are subjected to equal tensions. What will be the ratio of the frequencies of the notes emitted when each wire is sounding its fundamental?

The specific gravity of aluminium is 2·65 and that of steel is 7·8.

---

13. Prove that the velocity of the waves in a string is $\sqrt{T/\rho}$.

14. What is the frequency of a string whose length is 100 cm. and whose weight is 1 gm., when stretched by a weight of 20 kgm.?

15. If an addition of 25 lbs. to the tension of a string raises its pitch a fifth, what was the original tension?

16. Shew by a method similar to that of § 53 that the velocity of a particle at any moment is given by the equation

$$V = \frac{2\pi a v}{\lambda} \cos \frac{2\pi}{\lambda} (vt - x)$$

CHAPTER IV

1. What do you understand by longitudinal and transverse waves? A number of particles are arranged in a straight line at equal distances apart. Shew on a diagram the relative positions of the particles at a particular instant when (a) a longitudinal wave and (b) a transverse wave is passing along the row of particles.

2. Shew how the displacement at any instant of successive layers of air through which sound-waves are being propagated may be graphically represented. Indicate on your diagram the layers having maximum and minimum velocity respectively, and those at maximum and minimum pressure respectively.

3. Sound travels with a speed of 1120 ft. per sec. at 60° F. What are the wave-lengths of notes with frequencies 32 and 256? What is the frequency of a note whose wave-length is 1 in.?

QUESTIONS 285

4. State the law of variation of the intensity of sound at a point with the distance of the point from the source of sound, giving a general explanation of the cause of the change of intensity. Does the wave-length or the amplitude differ at different distances from the source?

5. State and shew by diagrams the way in which condensation is related to displacement in a progressive wave and in a stationary vibration in air.

6. Describe the motion of the air in two adjoining segments of a train of stationary vibrations.

7. What is a wave-front? How is it related to the direction of propagation of the sound?

8. State generally the nature of the reflection of sound at the closed and open end of a pipe respectively, and give the reason for the difference.

9. Give a general account of the distribution of energy in (*a*) a progressive wave and (*b*) a stationary vibration.

10. Shew that if sound travels along a tube which has a sudden change of bore at some point, there will be reflection of the sound at that point.

11. Shew that if curves similar to those of Fig. 41 be drawn for waves travelling towards the left, the convention as to the meaning of upward and downward ordinates remaining the same, the velocity curve will not occupy the position shewn in Fig. 41.

CHAPTER V

1. When a regiment of soldiers is marching behind a band, it is seen that the men are not all in step with each other but a kind of wave travels backwards along the column. Why is this?

2. A band is playing at the head of a procession 1080 ft. long, and the men step 128 paces to the minute exactly in time with the music as they hear it. Those in the rear are exactly in step with those in front. What is the velocity of sound?

3. Devise an experiment to shew that the velocity of sound in air varies with the temperature.

4. At what temperature is the velocity of sound in air 1200 ft. per sec., assuming that at 0° C. its velocity is 1090 ft. per sec.?

5. How could you shew experimentally that the velocity of sound in air is independent of the pressure?

QUESTIONS

6. Find the wave-length in hydrogen of a note whose frequency is 200. Assume that the velocity of sound in air is 1100 ft. per sec. and that the density of air is 14·4 times that of hydrogen.

7. A sound generated in water has a wave-length of 580 cm. in water. If the velocity of sound in water is 145,000 cm. per sec. and in air 34,000 cm. per sec., find the frequency and wave-length of the note heard by an observer in air.

8. Explain the nature of the error made by Newton in calculating the velocity of sound, and the nature of Laplace's correction.

9. The temperature of air through which sound-waves are propagated is supposed to be subjected to changes of an alternating character. Describe the nature of these changes and give an explanation of them.

10. The sound of a gun from a distant fort was heard at a certain house 20 sec. after the appearance of the flash. On another day the interval was 21 sec. What causes may have contributed to the difference?

11. Equal volumes of hydrogen and oxygen are mixed with each other. What is the ratio of the velocity of sound in the mixture to that in oxygen? Assume that the density of oxygen is 16 times that of hydrogen.

12. What evidence could you give that the velocity of sound is practically independent of the amplitude and frequency of the air vibrations?

13. A heavy piston moves without friction in a closed cylinder containing air and is set vibrating. Shew that the work done on the air in the cylinder when the piston describes its inward path is equal to the work done by the air when the piston describes its outward path, provided the compressions are either isothermal or adiabatic, but that the two quantities of work are not equal if there is a transference of heat between the air and the walls of the cylinder less than is required to make the compressions isothermal. Shew that in the latter case the vibrations of the piston will be damped and thence deduce Stokes' conclusion that the compressions and rarefactions in a sound-wave must be either isothermal or adiabatic.

CHAPTER VI

1. Explain the formation of echoes. A person standing between two high parallel walls makes a sharp sound. Account for the series of echoes he will hear. Consider the case where the observer is half way between the walls and the case where he is nearer to one wall than to the other.

QUESTIONS

2. Two men are equidistant from the face of a plane vertical cliff and are 1000 ft. apart. On one of them firing a pistol the other hears the echo one second after hearing the direct report. The velocity of sound being 1100 ft. per sec., find the distance of the men from the cliff. Would the interval between the two sounds be longer on a hot or a cold day?

3. A person standing at the end of a row of posts one foot apart makes a sharp sound which he hears reflected from each of the posts in succession. What is the frequency of the note resulting from the series of echoes?

4. A vibrating tuning-fork is placed in front of a wall. How could you find the positions of the nodes formed between the fork and the wall, and how could you use your observations for calculating the frequency of vibration of the fork? Why are the nodes sharpest near the wall?

5. Explain why a whisper at one point in a large room can sometimes be heard plainly at some other point.

6. A source of sound is situated under water. Give a diagram shewing the change in the direction of the rays when they emerge from the water.

7. Account for the difference between light and sound as regards the formation of shadows.

8. How is the apparent pitch of a sound affected by (1) motion of the source, (2) motion of the observer, (3) wind?

Two horns on a moving motor car sound a perfect fifth. Will a stationary observer hear the same or a different interval?

9. The note sounded by the horn of a motor car falls a whole tone in pitch as it passes a stationary observer. Shew that the car must be travelling about 45 miles per hour.

10. Devise an experiment for proving directly the laws of reflection of sound.

11. A train is approaching a hill from which a well-defined echo can be heard. The engine driver sounds his whistle. What will be the character of the echo heard (1) by the engine driver and (2) by an observer who is stationary?

12. A train approaches a stationary observer, the velocity of the train being one-twentieth the velocity of sound, and a sharp blast is blown with the whistle of the engine at equal intervals of a second. Find the interval between the successive blasts as heard by the observer.

13. Shew that in Example 1 the observer will hear simultaneously three series of equidistant echoes, the interval between the members of a series being the same in each of the three series.

14. A train whistles as it passes a stationary observer with velocity v. Find the interval between the notes heard by the observer when the train is approaching him and when it is leaving him, and shew that the total fall of pitch would be the same if the observer were in the train and the whistle were stationary.

CHAPTER VII

1. State what you understand by the term Interference. Mention several instances.

2. Describe the method of finding the velocity of sound by Seebeck's Tube, and mention any advantages the method possesses as compared with those carried out in the open air.

3. Shew that when two trains of similar waves cross each other at an angle there will be changes of velocity and displacement at the crossing point, but there may be no change of pressure.

4. Two tuning-forks have nearly the same pitch. Explain how to use the method of beats to find which is the higher and to find the difference between their frequencies.

5. Shew that the number of beats per second of two notes a semitone apart is different at different parts of the scale. Find the frequency of the beats when the lower of the two notes has frequencies 50, 250 and 1000 respectively.

6. What are Combination Tones and how are they produced? Find the frequencies of the first and second Difference Tones and of the first Summation Tone of two notes whose frequencies are 100 and 150.

7. Find the first difference tone for each of the consonant intervals within the limits of an octave.

8. Whilst the middle C of a harmonium is strongly sounded, the notes A, G, F and E from the octave below are successively sounded with it. What succession of notes will the first Difference Tones give?

9. If two organ pipes in exact unison be sounded at opposite ends of a large room, what will be heard by a man who walks from one of the pipes to the other?

10. If a vibrating fork is rapidly moved towards a wall, beats may be heard between the direct and reflected sounds. Account for them and calculate their frequency if the fork makes 512 vibrations per second and approaches the wall with a velocity of 300 cm. per second.

QUESTIONS 289

11. What is the velocity of sound in a gas in which two waves of length 1 and 1·01 metres respectively produce 10 beats in 3 seconds?

12. A wire stretched by a weight of 4 kilogrammes made 3 beats per second when sounded with a fork, and 5 beats per second with the same fork when the stretching weight was increased by 400 grammes. An addition to the stretching weight of 100 grammes made the beats slower in the first case and quicker in the second. Find the frequency of the fork.

CHAPTER VIII

1. Explain the meaning of the terms Free Vibration, Forced Vibration and Resonant Vibration. State in general terms the effect of a periodic force in producing vibrations in an elastic body.

2. A tuning-fork is struck smartly and held with its base on a sound-board. After the sound has died away the fork is again struck with the same intensity and held in the hand away from the sound-board. Which arrangement gives (1) the louder sound and (2) the longer duration of sound? Give reasons for your answer.

3. Why must two tuning-forks be very nearly in unison to shew resonance, whilst two strings on the same sound-box give resonance when they are only approximately in unison?

4. A fork making 256 vibrations per second is held over a tall narrow jar filled with water, and the water is allowed to run out gradually through a tap at the bottom until maximum resonance is obtained. What is the depth of the water level below the top of the jar? Find two other notes to which the same column of air would resound. (The correction for the open end may be neglected.)

5. A tube closed at one end is 10 inches long and 1 inch in diameter. Find the frequency of the note to which it resounds.

6. A tube 2 inches in diameter and open at both ends resounds to the note g. What is its length? ($c = 261$.)

7. Prove that the frequency of a Helmholtz Resonator is approximately inversely proportional to the square root of its volume. State in general terms how the pitch is affected by variations in the size and shape of the mouth.

8. Find the vibration ratio of the interval between two Resonators which are of the same size, but one of which has only a single opening while the other has three, the openings being all of the same size and shape. What would be the nature of the effect on the interval if the three openings were close together?

CHAPTER IX

1. State Helmholtz's Theory of the cause of the differences of quality of the notes of different instruments.

2. How would you prove that the note given by plucking a stretched string is not a simple tone? Describe and explain the differences of quality due to plucking at different points and with different instruments.

3. Describe three different methods by which you could shew that the note of a pianoforte contains a harmonic a twelfth above the fundamental. State how one or more of your methods would be interfered with if the pianoforte were badly out of tune.

4. How can an approximately pure tone be produced?

5. A key on a pianoforte is held down and the next adjacent octaves above and below are separately struck staccato. Describe and explain the result in the two cases.

6. Two strings are stretched side by side on a monochord and tuned to C and c. If either string is sounded and then silenced, the other will be found to be sounding c. Explain this.

7. Shew why the resonance of a massive body such as a tuning-fork, which, when set in vibration, goes on for some time, is a more precise indication of the existence in a sound of a component of definite frequency than the resonance of a body of less mass, such as a column of air.

8. What is meant by a Periodic Curve? State generally the nature of Fourier's analysis of such a curve. What is the application of the analysis to a complex musical note?

9. How does the mouth modify the quality of singing?

10. State the rival theories as to the nature of vowel sounds.

---

11. Compare the eye and the ear as regards their power of analysing complex vibrations. What are the limitations to the resolving power of the ear?

12. State Ohm's Law as to the physical basis of pure tones. What is the nature of the evidence on which it rests?

13. Give a general account of Helmholtz's method of synthesizing complex notes.

14. How did Helmholtz shew that the relative phases of the constituents of a complex note have no effect on the quality of the note?

CHAPTER X

1. Shew that the wave-length of the fundamental of a closed organ pipe is four times the length of the pipe.

2. Shew that if a closed organ pipe is so narrow that the correction for its open end may be neglected, the proper tones of the pipe will have frequencies in the ratio 1, 3, 5 etc.

3. Describe the motion of the air in an open organ pipe which is giving out its first overtone. What change will take place in the note and the mode of vibration, if a hole is bored in the pipe near its centre?

4. How do the motions of the particles of air within a sounding organ pipe differ from the motions of the particles of air outside the pipe by which the sound is transmitted to a distance?

5. Assuming that the velocity of sound in air is 1100 ft. per sec., find the approximate vibration number of an open pipe 4 ft. long, neglecting the correction for the open ends.

6. Explain why the note falls nearly but not quite an octave, when you close the open end of an open organ pipe.

7. Explain why the end of a reed pipe at which the reed is situated must be regarded as a closed end.

8. Describe and explain the various methods of tuning organ pipes.

9. How can the mode of vibration of the air in an open organ pipe be shewn experimentally?

10. An open pipe has a manometric capsule at its middle point, and another a quarter of its length from one end. The pipe is made to give its first four tones in succession. Describe the effect on the flames in each case.

11. Give a general statement of the effect of a change of temperature on the pitch of different classes of pipes.

12. Shew that two flue pipes of different pitches will rise in pitch by the same interval for a given rise of temperature.

13. A flue pipe gives the note C at 15°C. At what temperature will it give the note C♯, a semitone higher?

14. Discuss the effect of the correction for the open end on the relative frequencies of the proper tones of a pipe.

15. Explain why the note of a wide pipe contains fewer harmonics than the note of a narrow pipe.

16. Why are the higher harmonics of a reed pipe stronger than those of a flue pipe? Why do the reeds in an organ always have pipes associated with them?

17. A whistle gives a note whose frequency is 500 when blown with air at 20° C. When the same whistle is put in a furnace and again blown with air, it is found to give a note whose frequency is 1200. What is the temperature of the furnace?

CHAPTER XI

1. A rod is fixed at one end and has the other end free. Describe in general terms its possible types of vibration.

2. Explain the analogy between the longitudinal vibrations of rods and the vibrations of the air in organ pipes.

3. What are the advantages of a tuning-fork as a standard of pitch?

4. Explain the principle of Wheatstone's Kaleidophone.

5. How could you shew that when a plate is vibrating two adjacent sections separated by a nodal line are always moving in opposite directions?

---

6. Two tuning-forks have the same proportions, and are made of similar material. One of the forks is a fifth higher than the other. What is the ratio of their weights?

7. Explain why the frequency of the transverse vibrations of a rod fixed at one end is independent of the dimensions of the rod at right angles to the plane of vibration, but is not independent of the dimensions in the plane of vibration.

8. Shew that when a rod fixed at one end vibrates transversely, the nodes cannot be equidistant.

9. Shew that when a bell is giving its lowest proper tone, there must be both radial and tangential motion at the rim. How is this fact made use of in the ordinary method of making a wine glass sing by rubbing a wet finger round the rim?

CHAPTER XII

1. Mention the chief objections to the earlier methods of finding the velocity of sound in the open air by observing the interval between the flash and report of a cannon.

2. Describe some laboratory method of finding the velocity of sound in air.

3. Give a brief outline of two methods by which the velocity of sound in carbon dioxide could be found.

QUESTIONS 293

4. Describe some form of the Siren and explain how it could be used to find the frequency of vibration of an electrically maintained fork.

5. Describe a method by which you could determine as directly as possible the ratio of the frequencies of two tuning-forks.

6. How is the Tonometer used for finding the frequency of vibration of a tuning-fork?

7. Being given several forks of known frequencies and a resonance tube, what experiments would you make to shew that different notes travel in air with equal velocities?

8. State what change, if any, would be made in the dust figures in Kundt's experiment by a change in (1) the length and (2) the thickness of the vibrating rod.

9. State the advantages of the Lissajous' Figure method of comparing the frequencies of two forks as compared with the method of beats.

10. Two tuning-forks are known to be slightly more than a fifth apart. They are used for producing a Lissajous' Figure, and it is found that the figure completes its cycle of changes in 15 seconds. Assuming the frequency of the lower fork to be 261, find the frequency of the higher.

CHAPTER XIII

1. Explain the principle of the Phonograph.

2. A violin solo is reproduced on a phonograph, but the phonograph is run twice as fast as when the record was taken. Will the solo sound in tune?

3. Explain how the phonograph has been used to test the theories of the origin of vowel sounds.

4. Explain the action of the Bell Telephone.

5. Describe some form of Carbon Transmitter.

CHAPTER XIV

1. Describe and explain the sensations produced when the pitches of two notes originally in unison are gradually varied until they differ by a major third. In what way does the experiment depend on whether the notes are of high or of low pitch?

2. Two tuning-forks nearly an octave apart and free from overtones give beats when sounded together. What is the cause of the beats?

294   QUESTIONS

3. On what grounds do we suppose beats to be an important factor in our sensations of consonance and dissonance?

4. Use Helmholtz's method for comparing the consonance of a major third (4 : 5) and a major tenth (2 : 5).

5. A major sixth is played by an open and a stopped organ pipe. What difference in the effect will there be in the two cases when (1) the open pipe plays the lower note and (2) the stopped pipe plays the lower note?

6. A major third is given (1) on two clarinets, (2) on two hautboys and (3) one note on a clarinet and one on a hautboy. Explain the different degrees of dissonance.

7. What do you understand by a Consonant Triad? Find what consonant triads are possible within the compass of one octave.

8. Compare the consonances of the second inversion of a major triad and the second inversion of a minor triad, taking account of the Difference tones as well as the Harmonics.

CHAPTER XV

1. Explain how the existence of higher harmonic tones accompanying the fundamental in the sounds of a musical instrument assists the ear in estimating exactly a consonant interval.

2. An octave and a twelfth are both perfect consonances, yet a mistuned octave is worse than a mistuned twelfth. Why is this?

3. Two pure tones are to be tuned to a fifth. Show that the tuning is facilitated if the octave of the lower note is also sounded.

4. Why is it necessary that in any system of temperament the octaves must be true, whilst the minor thirds may be considerably different from true minor thirds?

5. What is the true diatonic scale and what are its advantages over other scales?

6. How could the monochord be used to tune eight notes on a pianoforte to a true diatonic scale?

7. Shew that if the scale of C on a pianoforte is tuned with true intonation, the note A so obtained will be out of tune, if used as the supertonic of the scale of G.

8. Why is it impracticable to have the musical intervals perfectly true on a pianoforte?

9. Find the interval between successive semitones in the equally tempered scale.

---

10. On the pianoforte twelve fifths make seven octaves. Find from this the error of a fifth.

11. Shew that an equal temperament minor third is about $\frac{8}{11}$ comma flat by using the fact that four equal temperament minor thirds make an octave.

CHAPTER XVI

1. Why do the strings of a pianoforte differ in thickness and length?

2. State generally how the quality of the note of a pianoforte is affected by (1) the point struck, (2) the shape of the hammer and (3) the hardness of the hammer.

3. Explain the method of producing the scale on a Brass Instrument with valves. If the valves are tuned correctly when used separately, the note produced when two valves are used together will be sharp. Why is this?

4. Explain how and why a rise of temperature affects the pitch of the wind instruments in the orchestra.

If the velocity of sound is 1120 ft. per sec. at 60° F. and 1140 ft. per sec. at 77° F., how much would a trumpet player have to alter the length of his instrument in order to keep to his original pitch, if the temperature of the instrument rose from 60° F. to 77° F.? (The tube of the trumpet in F is 6 ft. long.)

5. Would a given rise of temperature affect the pitch of a bass trombone and of a cornet to the same extent? Give reasons for your answer.

6. Why are the notes of a flute or clarinet put out of tune with each other, when the joints of the instrument are pulled out to lower its pitch?

7. Which of the instruments used in the orchestra can play in true intonation?

8. Which musical instruments have the same overtones as a stretched string?

---

9. Explain the principle and use of the Vibration Microscope.

10. Explain the action of a Violin Bow and describe the mode of vibration of the string of a Violin.

ANSWERS TO QUESTIONS

CHAP.

I. 7. The vibration ratio of a major sixth is obtained by doubling the lower of the two integers which give the vibration ratio of a minor third.

9. Difference of intervals is $\tfrac{10}{9}$. Sum of intervals is $\tfrac{5}{2}$. The two intervals are a fifth and a major sixth and their sum is a major third plus an octave or a major tenth.

II. 12. $\tau = 2\pi \sqrt{\dfrac{1000}{15 \times 981}} = 1\cdot 64$ sec. nearly.

15. Max. vel. $= a\omega = 10$, $\tau = \dfrac{2\pi}{\omega} = 2$, $\therefore\ a = \dfrac{10}{\pi}$.

16. Total energy $= \tfrac{1}{2} m \times$ (max. vel.)$^2 = \tfrac{1}{2} \times 20 \times 10^2 = 1000$ ergs.

17. If x is the length required, we have the relation
$x : x + 6\sqrt{3} = 1^2 : 2^2$ whence $x = 2\sqrt{3}$ in.
The arrangement described is known as Blackburn's Pendulum.

III. 8. The lengths of the segments are $54\tfrac{6}{11}$ cm., $27\tfrac{3}{11}$ cm. and $18\tfrac{2}{11}$ cm. respectively.

9. We have $100 = \dfrac{1}{2l} \sqrt{\dfrac{T}{\rho}}$ and $x = \dfrac{1}{4l} \sqrt{\dfrac{2T}{\rho}}$, whence $x = 50\sqrt{2}$.

10. $n_1 : n_2 = \dfrac{1}{2 \times 12} \sqrt{\dfrac{64}{\rho}} : \dfrac{1}{2 \times 15} \sqrt{\dfrac{36}{\rho}}$, whence $n_1 : n_2 = 5 : 3$ or the interval is a major sixth.

11. The mass of the string having the greater tension is double that of the other.

12. If n_1 is the frequency of the aluminium string and n_2 the frequency of the steel string, then $n_1 : n_2 = \sqrt{7\cdot 8} : \sqrt{2\cdot 65}$.

14. $n = \dfrac{1}{2 \times 100} \sqrt{\dfrac{20{,}000 \times 981}{\cdot 01}} = 221\cdot 5$.

15. $\dfrac{3}{2} = \dfrac{\sqrt{T+25}}{\sqrt{T}}$, whence $T = 20$ lbs.

IV. 3. Use the equation $v = n\lambda$. The wave-lengths are 35 ft. and $4\tfrac{3}{8}$ ft. The frequency is 13,440.

ANSWERS TO QUESTIONS 297

CHAP.

V. 2. The sound travels the whole length of the column in the time the men take to make 2 steps, i.e. it travels 1080 ft. in $\frac{60 \times 2}{128}$ sec.; whence $v = \frac{1080 \times 128}{60 \times 2} = 1152$ ft. per sec.

4. $\frac{1200}{1090} = \frac{\sqrt{x° + 273°}}{\sqrt{273°}}$, whence $x = 58°$ C.

6. 20·9 ft.

7. Frequency 250. Wave-length 136 cm.

11. 1·37 : 1.

VI. 2. The path taken by the echo must be 1100 ft. greater than the distance between the men, whence by geometry we find the distance of the men from the wall is 923 ft. The interval is greater on a cold day.

3. The period of the note heard is the time taken by the sound in travelling twice the distance between two adjacent posts. Taking the velocity of sound to be 1100 ft. per sec., the frequency is 550.

12. The interval is ·95 sec.

VII. 5. The ratio of the frequencies is in each case 15 : 16, whence the differences of the frequencies are 3·3, 16·7 and 66·7 respectively.

6. First Difference Tone 50. Second Difference Tones 50 and 100. Summation Tone 250.

10. If the person who carries the fork is the listener, he will hear 9·1 beats per sec. A stationary listener in the line of motion of the fork hears no beats when the fork is approaching him, and 9·1 per sec. after it has passed him.

11. 336¾ metres per sec.

12. If n is the frequency of the fork, n' that of the string with 4000 grm. and n'' that of the string with 4400 gm. we have $n' = n - 3$ and $n'' = n + 5$, whence

$$\frac{n-3}{n+5} = \frac{\sqrt{4000}}{\sqrt{4400}} \text{ and } n = 166·9.$$

VIII. 4. 1·07 ft. 256×3 and 256×5.

5. The effective length of the pipe is 10·3 in. Hence the wavelength is 41·2 in. and if the velocity of sound is taken as 1100 ft. per sec. the frequency is 320·4.

6. 15·6 in.

8. $n : n' = \sqrt{1} : \sqrt{3} = 1 : 1·73$.

X. 5. 137·5.   13. 54·6° C.   17. 1416° C.

XI. 6. 27 : 8.   XII. 10. 391·57.   XVI. 4. 1⅗ in.

INDEX

The figures refer to the pages

Adiabatic coefficient of elasticity of a gas, 83
Aircraft, detection of, 274
Air-waves, general description of propagation, 59
Amplitude defined, 25
Analysis of complex vibrations, 154
Antinode defined, 49
Appun's tonometer, 202
Associated curves, 64

Bassoon, 262
Beating reeds, 177
Beats, general explanation, 121
——, pitch of note heard, 123
——, experimental illustrations, 123
——, due to combination tones, 127, 227
——, measurement of pitch by, 201
——, effect on the ear, 225
——, effect on consonance of pure tones, 225
Bell-jar experiment, 2
Bells, vibration of, 190
——, overtones of, 192
——, tuning of, 193
Bell telephone, 219
Binaural hearing, 274
Blaikley, on length of wind instruments, 86
——, on correction for open end, 174
——, on velocity of sound, 209
Brass instruments, 264; shape of, 264; crooks of, 264
Bravais and Martin, on velocity of sound at high altitudes, 206
Bugle, 265

Cagniard de le Tour's siren, 196
Carbon transmitter, 222
Chladni's figures, 189
Clarinet, 232, 259
Clément and Desormes, on ratio of specific heats, 84
Closed end of tube, reflection at, 69
Closed organ pipe, period of, 166; overtones of, 169
Colladon and Sturm, on velocity of sound in water, 206
Combination tones, 124; theories of origin, 125; method of finding pitch, 126; formed by harmonics, 233
Common chord, 236
Comparison of intensities of sounds, 214
Complex vibrations, analysis of, 154; synthesis of, 158
Composition of simple harmonic motion with uniform motion, 27
—— of two simple harmonic motions, 28, 30, 32
—— of harmonic vibrations produced optically, 34, 185, 252
Concentration of sound by spherical mirror, 93; by walls of room, 95
Condensation in air-waves, 62
——, absolute, in organ pipes, 213
Conical pipes, overtones of, 176
Conjugate foci, 94
Consonance, 224
——, Helmholtz's theory of, 227
—— of intervals of pure tones, 227
—— of intervals of complex tones, 228
—— of notes which have not the full series of harmonics, 232
—— of triads, 236

INDEX

The figures refer to the pages

Consonant intervals, defined, 9; relative smoothness of, 224; analysis of relative smoothness of, 228; effect of widening the intervals by an octave, 230
Consonant triads, defined, 233; derivation of, 235; effect of difference tones, 237
Cor de chasse, 265
Cornet, 270
Correction for open end, 167; method of finding, 173; effect on overtones, 174; effect on quality, 175
Crooks of brass instruments, 269
Crova's disc, 65
Curves of displacement of vibrating particles, 25; relation to velocity curve, 27; by Lissajous' Figures, 32; in transverse wave motion, 35, 71; in stationary vibrations, 48; in air-waves, 64, 71, 74; in beating notes, 122; in complex vibrations, 148; in organ pipes, 168

Definition of intervals, 239; of pure tones, 239; of complex notes, 240
Diatonic scale, defined, 10; derivation and advantages of, 242; defects of, 244
Difference of phase defined, 25
Dimensions, method of, 193
Disc siren, 8
Displacement. *See* Curves of displacement
Dissonance due to beats, 226
Dominant, 244
Doppler's principle, 103; applied to light, 107
Drums, 193, 273

Ear, described, 143
——, resolving power of, 145
——, limitations of, 146
Ear trumpet, 69
Echoes from plane surfaces, 92
—— from palings, 97

Elastic deformation, relation to deforming force, 15
Elastic vibrations, general description of, 18; isochronism of, 19; calculation of period of, 23
Elasticity, nature and limitations of, 13
——, imperfect, 14
—— of liquids, 15, 17
—— of gases, 15, 57
——, coefficient of, 24, 82, 83
Energy transmitted along train of stationary waves, 50
—— of air-waves, 68
Equal temperament, 247; errors of intervals in, 248
Equal temperament semitone, vibration ratio of, 248

Flue-pipes, described, 165
——, period of, 166
——, overtones of, 167
——, methods of tuning, 172
——, conical, 176
Flute, 256
Fog signals, zones of silence near, 120
Forced vibrations, 129; amplitude of, 137; phase of, 139; initial stages of, 139; used in musical instruments, 140
Fourier's theorem, 150; applications of, 152
Free reeds, 177
Free vibrations, defined, 128
French horn, 265, 271
Frequency, defined, 7
Frets, 255
Fundamental defined, 55

Geophone, 276
Graphic method of measuring pitch, 199

Hand horn, 265
Harmonic constituents of notes of musical instruments, 163
Harmonic series, 55, 151, 224

The figures refer to the pages

Harmonics, defined, 55, 153
———, given by Fourier's theorem, 151
———, quality dependent on, 154
——— in the voice, 162
——— of organ pipes, 175
———, influence on consonance, 228
———, influence on definition of intervals, 240
——— of violin, 164, 255
——— of flute, 164, 259
——— of clarinet, 164, 261
——— of hautboy, 164, 262
Harmonium, 177
Harmonograph, 28
Harp, 252
Hautboy, 232, 261
Helmholtz, resonators, 131; *see also* Resonators
———, theory of quality, 154
———, analysis of complex vibrations, 155
———, synthesis of complex vibrations, 158
———, theory of consonance, 227
———, on the vibrations of violin string, 252
Hooke's Law, 16
Hughes's microphone, 221
Hunning's transmitter, 222

Intensity of sound, variation with distance, 69; near interfering sources, 115; absolute measurements of, 211
Interference, meaning of the term, 109
——— near two sources, 115
——— shewn by branched tube, 116; by tuning fork, 117; by Seebeck's tube, 118; near fog signal, 120
Intervals, 8
———, measurement of, 9
———, consonant, 9
———, sum of, 10
———, difference of, 10
———, definition of, 239

Intervals. *See also* Consonant intervals
Inversions of triads, 235
Isochronism, 19
Isothermal coefficient of elasticity, 82

Kaleidophone, 185
Key bugle, 266
Kohlrausch, on sensation of pitch, 203
König's manometric capsule, 171
Krakatoa eruption, 206
Kundt's method of measuring velocity of sound, 209

Laplace's correction of Newton's calculation of the velocity of sound, 82
Leading note, 244
Limits of audibility, 7
——— of elasticity, 13
Linear dimensions, relation to pitch, 194
Lissajous' figures, 32; optical method of producing, 34; produced by kaleidophone, 186; used for measuring differences of frequency, 202
Location of sound in air, 274
——— of sound underground, 275
——— of sound under water, 276
——— of gun by sound, 277
Longitudinal vibrations of air particles, 60
Longitudinal waves, 57; illustrated by spiral spring, 58; properties of, 61, 64; condensation in, 62; shewn by Crova's disc, 65; velocity of, 79; superposition of, 109
Loudness of sound, 5

Manometric capsule, 171
Mayer, on comparison of intensities, 214
Mean tone temperament, 246
Mediant, 244

INDEX

The figures refer to the pages

Melde's experiment, 142
Membranes, vibrations of, 193
Microphone, 221
Mining, located by sound, 275
Mixture stops, 242
Modulation, 244
Monochord described, 51
——, resonance shewn by, 131
—— used for measuring pitch, 198
Mouth of organ pipe, action of, 165
Musical box, 186
Musical echo from palings, 97
Musical instruments, described, 250; qualities of notes of, 163
Musical notes, characteristics of, 3, 5

Natural modes of vibration, 153; *see also* Overtones
Newton's calculation of velocity of sound, 81; Laplace's correction of, 82
Nodes defined, 49
Noise, 3
Nomenclature of stationary vibrations, 49
—— of complex notes, 153
Non-harmonic force, effect in producing vibrations, 129, 158
Non-harmonic waves, 39
Note defined, 153

Ocarina, 241, 259
Ohm's Law, 125, 147
Open end of tube, reflection at, 75; correction for, 78, 167, 173
Open organ pipe, period of, 166; overtones of, 168
Ophicleide, 266
Organ pipes, flue, 165; reed, 176; amplitude of vibration of air in, 213
Overtones, defined, 55
—— of strings, 55
——, harmonic and inharmonic, 153
—— of organ pipes, 167
—— of rods, 182

Overtones of tuning-forks, 187
—— of plates, 189
—— of bells, 192
—— of clarinet, 260
—— of hautboy, 262
—— of brass instruments, 264

Partials, defined, 56
Periodic curves, 149
Personal equation, 205
Phase, defined, 25
—— of forced vibrations, 139
——, effect on quality, 159
Phonograph, described, 216
—— used to test vowel theories, 218
Pianoforte, 250
Pipes of variable bore, 76
——, velocity of sound in, 207
——, *see also* Organ pipes
Pitch, measured by frequency, 6
—— notation, 11
—— of note given out from moving source, 104
—— of note heard by moving listener, 105
——, dependence on dimensions of sounding body, 194
——, measurement by siren, 196; by monochord, 198; by graphic method, 199; by tonometer, 200
——, sensation of, 203
——, standards of, 263
Plane waves, constancy of intensity of, 69
Plates, vibrations of, 189
Potential energy of deformed body, 18
Proper modes of vibration, 153
Pure tones, consonance of, 225, 227

Quality, characterized by wave form, 12
——, meaning of the term, 145
——, Helmholtz's theory of, 154
——, dependence on phase of constituents, 159

The figures refer to the pages

Ratio of specific heats, 84; determined by Kundt's method, 211
Rayleigh, on correction for open end, 173
———, on absolute measurement of intensity, 211
———, on comparison of intensities, 214
Reciprocal firing, 204
Reed pipes, 176
———, method of tuning, 179
Reeds, free and beating, 177
———, effect on pitch of pipes, 178
Reflection of waves at end of string, 45
—— of waves at closed end of pipe, 69
—— of waves at open end of pipe, 75
—— at surface of changing density, 77
—— of spherical waves, 90
—— of sound by spherical mirrors, 93
———, total internal, 99
Refraction, 99
Regnault, on measurement of sound in open air, 205; in pipes, 207
Resonance, 128
———, effect of mistuning on, 137
Resonance box of fork, 136
Resonant vibrations, of pendulum, 128; instances of, 130; shewn by monochord, 131
Resonators, Helmholtz's, 131
———, nature of vibrations, 132
———, pitch of, 133
———, conductivity of mouth of, 134
—— with several mouths, 135
Restitution forces, 17
Rods, longitudinal vibrations of, 181
———, transverse vibrations of, 183
Rücker and Edser, on combination tones, 126

Savart's toothed wheel, 6
Saxhorn, 272
Saxophone, 261

Scheibler's tonometer, 200
Seebeck's tube, 118, 208
Sensitive flame, 94
Simple harmonic vibrations, 21; geometrical illustration of, 22; period of, 23
Sine curve, 25
Slide of trombone, 267
Sound, simpler properties of, 1
—— vibrations superposed on molecular motions, 65
—— propagated at right angles to wave-front, 68
—— images, 92
—— shadows, 102
Sound ranging, 277
Soundboard of musical instruments, 140
Speaker key, 260
Speech, 160
Spherical waves, general description of, 67; variation of intensity with distance, 69; superposition of, 113
Spiral spring, law of stretching of, 15; illustrating longitudinal waves, 58; illustrating overtones of pipes, 170
Standards of pitch, 263
Stationary vibrations, of string, 48; experimental demonstration of, 49, 97; nomenclature, 49; in closed tube, 70; properties of, 72; compared with progressive waves, 75; in open tube, 77; effect on ear, 111
Stretched string, waves on, 41; velocity of waves on, 42; superposition of waves on, 44, 47; frequency of, 50; laws of vibration of, 51; experimental test of laws, 52; nodes on, 54; effect of imperfect flexibility, 56
Submediant, 244
Superposition of waves on string, 44, 47
—— of trains of waves, 109
—— of oblique trains of waves, 112

The figures refer to the pages

Superposition of spherical waves, 113
——, limits of law of, 124
Supertonic, 244
Synthesis of complex notes, 158

Telephone, 218
Temperament, 246
——, mean tone, 246
——, equal, 247
Temperature, effect on velocity of sound, 86
——, effect on direction of propagation, 101
——, effect on organ pipes, 179
——, effect on pitch of brass instruments, 272
Threlfall and Adair, on velocity of sound in water, 207
Tone, definition of, 153
Tonic relationship, 241
Tonometer, 200
Töpler and Boltzmann, on amplitude of vibration of air, 213
Total internal reflection of sound, 99
Transmitter, 222
Transverse waves, 35; velocity of, 37, 42; equation of, 40; on a string, 41; superposition of, 44, 47; not possible in a gas, 57
Triads. *See* Consonant triads
Triangle, 188
Trombone, 266
Trumpet, 270
Tuba, 272
Tube, reflection at closed end, 69
——, reflection at open end, 75
—— of variable bore, 76
Tuning an elastic body, 24
—— flue pipes, 172
—— reed pipes, 179
—— wood wind, 262
—— brass instruments, 272
—— drums, 273
Tuning-fork, on resonance box, 136
——, electrically maintained, 155
——, general description, 187
——, effect of temperature on, 187

Tuning-fork, overtones of, 187
Two adjacent sources of sound, 113

Valves of brass instruments, 268; defects of, 269
Velocity of particles, relation to displacement, 27
Velocity of sound, nearly independent of amplitude, 85; independent of pitch and pressure, 85; in different gases, 85; effect of temperature, 86; in mixtures of gases, 87; in liquids, 88, 206, 207; in solids, 88; in open air, 204; at high altitudes, 206; in pipes, 207
Velocity of sound measured, by echo, 93; by resonance tube, 208; by Seebeck's tube, 208; by organ pipes, 209; by Kundt's tube, 209
Velocity of transverse waves, 37; relation between velocity and wave-length, 38
Velocity of elastic waves of any type, 81
Ventral segment, defined, 49
Vibration microscope, 252
Vibrations. *See* Elastic vibrations
Violin, 252
——, motion of string of, 254
——, action of bow, 254
——, quality of, 255
Viscous liquids, 14
Vocal organs, 161
Vowels, 160
Vowel theories, 162, 218

Wave-front, 67
Waves. *See* Longitudinal waves *and* Transverse waves
Wheatstone's kaleidophone, 185
Whispering gallery, 96
Wind, effect on direction of propagation of sound, 100
——, effect on pitch, 107
——, effect on velocity measurement in open air, 204